The Life of t]

In this classic of natural history, ⸻ ⸻ ⸻ ⸻ derstand the sometimes curious habits of our red-breasted garden companions. With a mixture of scientific objectivity, easy speculation and friendly enthusiasm, each chapter considers a different aspect of the robin's behaviour. Topics include song, territory, courtship, migration and the robin's raising of its young.

Lack's own findings are supplemented throughout with references to other scientific studies and to observations of lay ornithologists who have written to him. Meanwhile, Lack's references to the robin's numerous appearances throughout English literature reveal Britain's long relationship with our newly voted national bird.

Unavailable for many years, this classic work is introduced by one of today's best-known ornithologists, David Lindo, Urban Birder and organiser of the 2015 vote. Other additions are provided by the author's son, Peter Lack, and by the doyen of robin studies today, David Harper, who describes recent advances in robin studies.

The book is illustrated by Robert Gillmor, recipient of the RSPB medal, and completed by Lack's bibliography and a fully updated index.

About the authors

David Lack (1910-1973) was a British evolutionary biologist who began his scientific observations while working as a schoolteacher at Dartington Hall. He made major contributions to ornithology, ecology and ethology, and his pioneering life-studies helped ornithology to become respected as a serious science.

David Lindo – the Urban Birder – is a naturalist, writer, broadcaster and educationalist. He has contributed to many shows and publications, including *Springwatch*, *BBC Wildlife Magazine* and the RSPB's *Birds*. He now travels the world teaching people about the amazing urban wildlife on their doorstep.

Peter Lack is David Lack's son. Based at the British Trust for Ornithology, he has published extensively about birds, with particular interests in birds in winter, migration strategies and African birds.

David Harper is senior lecturer in evolutionary biology at the University of Sussex. In addition to lecturing, he researches the behavioural ecology of passerine birds, especially robins.

Robert Gillmor is an ornithologist, artist, author and editor. He is perhaps best known for his covers for the Collins *New Naturalist* series. He is also a founding member of the Society of Wildlife Artists, and a recipient of both the RSPB medal and an MBE.

THE LIFE OF THE ROBIN

The Life of the Robin

DAVID LACK

Introduction by
David Lindo

Postscripts by David Harper
and Peter Lack

Illustrated by
Robert Gillmor

PALLAS ATHENE

Contents

winter quarters – Fighting on board ship – Redstart changing into robin – Swallows hibernating – Distances shifted by adult and juvenile robins.

The age to which birds can live in captivity and in the wild – The percentage of robins dying each year and the expectation of life as calculated from the ringing returns – Comparison with human life-tables – Number of young raised per pair per year.

The moral Bunyan – Normal food – Beneficial to man – Feeding methods – Pellets – Desire for fat – Effect of hard winters – Robins make cats vomit – Enemies – Parasites – As food and medicine for man – Huge annual mortality – Control of population.

Shape of robin's territory – Early references to territory – Value in pair-formation – Discussion of optimum spacing, food territories, attacks on food competitors, size of territory and factors determining it, and equal spacing of birds which do not feed in their territories – The autumn territory of the robin – Sexual behaviour in autumn – Male behaviour by female robin – Autumn territory and migration.

Abnormal behaviour throws light on the normal – Manner of experiments – Robins attack a stuffed specimen – Individual differences in fierceness and way of attack – Waning of fierceness with repetition – Alleged cannibalism – Experiments with parts of specimens – Courtship of the specimen – Attacking empty air – The robin's world.

Attacks on other species – Recognition a bad term – Lorenz's views on signals – Robin's attacking behaviour partially but not completely divisible into three parts each with its own signal – Lorenz's views probably over-simplified – Attack on the mate – Attack only in the territory – Essential to consider the internal state – Bird behaviour complex.

Text Figures

Introduction

What do you know about robins? Is it the same happy robin that eagerly greets you at your back door every time you nip out for a spot of gardening?

Those questions and many more intrigued me as a kid. I used to stand staring out of my bedroom window at the antics of the robin in my garden. I remember being shocked when I first learnt about our humble robin's violent *alter ego*. How could such a cute looking bird have such a foul temper?

I was equally horrified to find out that they were not immortal either. Yes, the robin you so adoringly called yours could have been a succession of similar looking birds over the course of years. And Lord knows, the bird born in your garden shed during the summer could end up spending its Christmas soaking up the winter sun in Spain and not shivering in the snow perched on a spade. Indeed, over the last fifty years we have been learning a lot about the life and times of the robin. All of that knowledge stems from the work of one man: David Lack. Reading through the pages of this book it is clear to see that David was ahead of his time and a total innovator in the techniques that he used to study his chosen subject. He borrowed the ideas and practices that went before him to help shape the exhaustive research that he carried out over the space of four years. I love the fact that David was not an out-and-out scientist. He was a biology teacher, an amateur, who took it upon himself to study a bird that had always captivated him. This work is a shining example of how useful a position citizen science occupies within the world of academia.

The robin has always had a role in the collective British psyche – so much so that in the 60's it was anointed as the nation's favourite bird. This announcement was made not as a result of public opinion; rather it was decreed in an office somewhere in London by a bunch of men in crusty suits. I had often wondered just how deep was Britain's love affair with the robin, so in 2014 I launched a campaign to get the country to decide once and for all on the bird best placed to represent all that Britain stands for. The first round of voting focused on a long list of sixty iconic species. Aiming at birders and those interested in nature, I asked voters to plump for up to six species that they fancied. The robin won that round by a country mile. The second round featured the ten most popular species from the first round. The robin's contenders included the blackbird, wren, blue tit, kingfisher, mute swan, puffin, red kite, barn owl and a rank outsider, the hen harrier.

The timing of the bird vote was crucial as it ran for two months alongside the other national election that was going on at the time. The final day of voting occurred on May 7th 2016, the date of the General Election, and the date that thousands of children across the land also cast their vote for Britain's National Bird. The electorate turned out in their droves and by midnight on May 7th nearly 250,000 people had had their say. Needless to say, the trusty robin won by a landslide, commanding over 30% of the vote.

The thing that I learnt from the experience was that many people already thought that the robin was Britain's National Bird. But many were as shocked as I was as a kid to learn of the highly territorial and pugnacious nature of their newly appointed avian emblem. I think that David Lack would have been proud to know that the robin is still as popular as it was back in his day. His words in this book are just as informative and surprising now as they were then, and his language just as engaging and approachable.

It has been an honour for me to contribute to this milestone in ornithology and I hope that David Lack will go on to inspire new generations of nature lovers, birders, ornithologists and, most importantly, robin lovers for years to come.

David Lindo, The Urban Birder, Spring 2016

Preface

The robin is the most popular British bird, perhaps the most popular European bird. Literature abounds with allusions to it of every kind, varying from the purely mythical – 'A Robbyn readbreast, finding the dead body of a Man or Woman, wyll cover the face of the same with Mosse. And as some holdes opinion, he wyll cover also the whole body' (Thomas Lupton, 1579) – to the eminently practical – 'This amiable little songster is eaten roasted with bread crumbs' (French recipe). The present book is concerned with the real life of the robin, which may be found more curious than the legends. Nor is it altogether without topical interest to consider a bird whose active life is devoted almost wholly to fighting, even though the robin is so inhuman as to achieve its victories without bloodshed.

This account is based on four years' observations which, to quote Gilbert White, 'are, I trust, true in the whole, though I do not pretend to say that they are perfectly void of mistake, or that a more nice observer might not make many additions, since subjects of this kind are inexhaustible.' Before this study was started in January 1935, the robin was credited with being far better known than any British bird. Nevertheless a considerable number of new, and in some cases startling, facts have come to light, and obviously much more has yet to be discovered. The same holds for all the common British birds. The pleasure of studying birds and the pleasure of finding out new things can be combined at small cost in money, though more in time, by anyone so inclined. Further, the study of bird behaviour is at so early a stage that it can be described in simple language without loss in precision, and this I have tried to do throughout.

Most of the robin's behaviour is described in general terms, but only where the action concerned has been seen on at least six occasions. For this there are two reasons. First, though in birds the main patterns of behaviour are the same for all the members of one species or subspecies, individual differences exist and are by no means negligible. Secondly, birds act much more quickly than human beings, and it is often difficult to see the whole of what takes place. Subconsciously the mind of the observer fills in the points which the eye misses, so that the bird's behaviour tends to conform with his preconception of what it is doing. This can lead to serious errors both of fact and in interpretation. The only check is repeated observation, and it is particularly helpful to see the same action in several different individuals, as they may perform it somewhat differently.

Any study such as the present must constantly refer to the work of others. Indeed I hope that a subsidiary value of this book may be that it provides a background summary of much of the modern work on bird behaviour. I have also, where relevant, included quotations from early works on natural history, works which are less generally known than they deserve. To avoid encumbering the text, actual references are detailed in a special appendix (p. 249), and readers wishing to check the source of any observation not my own should turn up the appropriate chapter and page number in this list.

In addition to these writers, I am personally indebted to a great many people for help in trapping birds, building aviaries, contributing facts, formulating ideas, checking references, loaning photographs, and criticizing the manuscript. The names of these numerous willing helpers make so long a list that I have decided to omit them, and I hope they will appreciate that their anonymity in no way reflects ingratitude on the part of the author.

For the benefit of American and Dominion readers, it should perhaps be noted that the subject of this book is the English robin, whose scientific name is *Erithacus rubecula*, and that this bird is not the same as, nor a particularly close relation of, the various red-breasted birds which have been called robins in other parts of the English-speaking world. The English robin is a bird rather smaller than a sparrow, in build between a thrush and a warbler, uniform brown on the upper parts, with an orange-red breast and white abdomen. It is widely distributed throughout the woodlands of Europe, and in Britain is also a familiar garden bird.

In wishing the reader a pleasant and profitable time, I cannot do better than to conclude with Thomas Coryat's oration to Prince Henry on presenting his book called *Crudities* to him in 1611: 'In the meane time, receiue into your indulgent hand (I most humbly beseech your Highnesse) this tender feathered Redbreast. (Because the Booke was bound in Crimson Velvet.) Let his Cage be Your Highnesse studie, his pearch your Princely hand, by the support whereof, hee may learne to chirp and sing so lowde, that the sweetnesse of his notes may yeeld a delectable resonancie.'

London, October 1941

Preface to the Fourth Edition

The first edition of *The Life of the Robin*, which appeared in 1943, was based on field work carried out at Dartington in South Devon between 1934 and 1938. An illness in March 1945 enabled me during convalescence to return briefly to Dartington that spring, also to make some more observations at nests elsewhere, and to ask other observers for their nest records. This new material, with further observations on food, was included in the second edition of 1946, in which parts of the text were also revised and rearranged. In the autumn of 1945 I came to live in Oxford, and there studied robins for another three years, especially their nesting and ecology, this work being included in a postscript to the third (Pelican) edition of 1953. The text of the present edition is that of the third except for a few small amendments, notably on the significance of courtship-feeding and of the colour of the eggs, and with a new Chapter 16 on forest robins, which includes material from the postscript to the third edition. I have also added a few observations on the behaviour of robins published by others between 1945 and 1964, so the book is still a full account of the English robin. But I have omitted recent work in museums on racial variations in the robin, and at British bird observatories on migrant Continental robins, which would have introduced problems remote from those discussed here. It also proved impossible, without destroying the balance of the book, to include new observations on other kinds of birds, or new theories of behaviour. Finally there is a fresh set of illustrations, undertaken by Robert Gillmor with his usual felicity.

As I noted in the preface to the second edition, while this book is based on original observations, it owes an immense debt to others – not merely to the great naturalists, which is obvious enough, but to the many observers of small incidents, published in notes to journals of natural history or sent in private correspondence. The naturalist does not work in isolation, but has many helpers, and while the contributions of any one of us are small, the united contributions form a considerable mass, which, even so, is negligible beside what has yet to be discovered.

Seeing that the book had by then justified a paperback edition, I concluded the preface to the third edition by quoting a caustic comment, made rather over a century ago by the distinguished Scottish ornithologist William MacGillivray: 'I have heard,' said he, 'of a closet naturalist who, slighting the labours of a brother in the field, alleged that he could pen a volume on the robin; but surely, if confined to the subject and without the aid of fable, it would prove a duller book than *Robinson Crusoe*.'

For this fourth edition, it seems fitting to end with a tribute to the two ornithologists, both now dead, to whom it owes most. One was J. H. Burkitt (1870-1959), a civil engineer by profession and county surveyor for Fermanagh in Northern Ireland, who was the first man to study robins intensively, indeed the first man to mark any population of wild animals individually, so that he unawares initiated a technical revolution in field studies. He wrote to me, after reading the first edition, that he did not start looking at birds until the age of thirty-seven, when he had no bird-books and no bird friends. Apart from his strikingly original work on the robin, begun after the age of fifty, he published almost nothing, perhaps because: 'When I was doing the robin, I had pricks of conscience that I was really more interested in the created than the Creator.' A religious, humble, and able man, he was immensely surprised at the great interest which his researches evoked, and he spent his declining years reading his Bible and cultivating his garden.

The other man was H. F. Witherby (1873-1943), formerly senior partner in the house which published this book, and he discussed the original manuscript with me in his home, Gracious Pond Farm, shortly before he died. He was the leading British ornithologist of his time, founder of the magazine *British Birds* in 1907 and of the national bird-ringing scheme in 1909, and senior author of the standard work on British birds, the *Practical Handbook* of 1919-24 and its greater successor the *Handbook* of 1938-41. Accurate, humble, kindly, and a prolific worker, he did more than anyone else to raise the standards of amateur ornithology. I might add that the Algerian race of the robin, *Erithacus rubecula witherbyi*, is called after him, and, more important, that each of my robins bore on one leg what was then generally known as a 'Witherby ring'.

Oxford, May 1965

Grateful acknowledgement is made to Mrs W. H. Davies and Jonathan Cape Ltd for permission to quote two lines from the poem 'Robin Redbreast' from the *Collected Poems of W. H. Davies*, and to Robert Gillmor for the entirely new set of illustrations which replace those of earlier editions.

Note to this edition

This edition is a reprinting of the final edition overseen by David Lack, in 1965. As well as the new introduction by David Lindo, a chapter outlining more recent developments has been contributed by David Harper (p. 221), and an essay about the history of the writing, publication and influence of the book, by David Lack's son Peter, is printed on p. 233. For the sake of clarity, the original references and notes to the main text have been kept together; they are printed, and followed by the notes to the two later essays. We are very pleased to have been allowed to reproduce Robert Gillmor's fine illustrations for the 1965 edition, and fifty years later he has kindly created a new cover specially for this edition.

Dedicated
to all those robins
who patiently bore my rings
and permitted my intrusions
into the intimacies of
their lives

The world was made to be inhabited by beasts, but studied and contemplated by man: 'tis the debt of our reason we owe unto God, and the homage we pay for not being beasts. Without this, the world is still as though it had not been, or as it was before the sixth day, when as yet there was not a creature that could conceive or say there was a world. The wisdom of God receives small honour from those vulgar heads that rudely stare about and with a gross rusticity admire His works. Those truly magnify Him whose judicious enquiry into His acts and deliberate research into His creatures return the duty of a devout and learned admiration.

SIR THOMAS BROWNE: *Religio Medici* (1643)

MR. ASTERIAS: A morbid, withering, deadly, antisocial sirocco, loaded with moral and political despair, breathes through all the groves and valleys of the modern Parnassus; while science moves on in the calm dignity of its course, affording to youth delights equally pure and vivid, to maturity, calm and grateful occupation, to old age, the most pleasing recollections and inexhaustible materials of agreeable and salutary reflection; and, while its votary enjoys the disinterested pleasure of enlarging the intellect and increasing the comforts of society, he is himself independent of the caprices of human intercourse and the accidents of human fortune. Nature is his great and inexhaustible treasure. His days are always too short for his enjoyment: *ennui* is a stranger to his door. At peace with the world and with his own mind, he suffices to himself, makes all around him happy, and the close of his pleasing and beneficial existence is the evening of a beautiful day.

THE HONOURABLE MR. LISTLESS: Really I should like very well to lead such a life myself, but the exertion would be too much for me. Besides, I have been at college.

THOMAS LOVE PEACOCK: *Nightmare Abbey* (1817)

So we rode around the park until quite late talking and philosophizing quite a lot. And I finally told him I thought, after all, that bird life was the highest form of civilization. . . Gerry says he has never seen a girl of my personal appearance with so many brains.

ANITA LOOS: *Gentlemen Prefer Blondes* (1926)

I

My Robins

Will you not look twice now before picking up crumbs scattered by strangers? No man in the world is so dreadful as the bird catcher.

MRS SARAH TRIMMER: *A History of the Robin*
(1786)

This study was carried out in a region of woodland, orchards, quarries and fields at Dartington in South Devon. The first stage was to mark all the wild robins of the neighbourhood in such a way that they could easily be distinguished from each other without disturbing them. For this purpose, coloured rings on the legs proved much the most suitable method.

Two and three hundred years ago, and even in Victorian times, there was a widespread knowledge among country people of how to catch small birds, which were used both for eating and as cage birds. The robin was popular as a cage-bird (cf. Blake's famous lines), and was even eaten, Thomas Muffett writing in *Health's Improvement* in 1595, 'Robin-redbreast is esteemed a light and good meat'. But with the rise in the standard of living the eating of small birds gradually died out in England, and the growth of popular sentiment against cage birds has finally suppressed the catching of small birds as a livelihood. As a result, when the ringing of birds for scientific purposes commenced in the present century, the old bird-catching lore had been practically forgotten. For catching robins this mattered little. 'The way of taking a Robin-red-breast is so easie and common, that every Boy knows how to take him in a Pitfall,' wrote Nicholas Cox in the *Gentleman's Recreation* in 1674, and a century later Buffon* was writing, 'It is always the first bird that is caught by the decoy.'

The trap found most convenient in the present investigation was a small version of the American house-trap. It consists simply of four walls and a roof of wire netting, five feet high with a base of three feet six inches square. It could be carried conveniently by two people, and was placed on flat ground, food being scattered inside. The bird enters by a small funnel at ground level, and, once in, it hardly ever finds its way out again. This is apparently because, when alarmed, the habit of a bird is to fly upwards, hence it does not notice the exit at ground level. Of all the birds which entered the house-traps, only house-sparrows and an occasional blue tit were able to go in and out regularly without getting caught; house-sparrows are perhaps more intelligent than most birds.

---

* References to observations other than my own are detailed in a special appendix, p. 249; there is normally no indication in the text.

In winter robins are easy to trap, for they soon come down to investigate any strange object in the territory. If they were human they would be described as curious, but it is difficult to know just what this means in a bird's mind. The habit does not seem to be due wholly to the robins' intensive search for food, for they come down to investigate anything unusual in their territories, including objects which appear to have no possible connexion with food. In the eighteenth century Buffon noted that robins would follow travellers through the forests, and perhaps it was this which gave rise to the legend that

> Covering with moss the dead's unclosed eye,
> The little redbreast teacheth charitie

as Drayton puts it. Compare also, Shakespeare's 'ruddock with charitable bill', and, most famous of all:

> Thus wandered these poor innocents
> Till death did end their grief;
> In one another's arms they died
> As wanting due relief.
> No burial these pretty babes
> Of any man receives
> Till Robin Redbreast piously
> Did cover them with leaves.

For trapping robins, the more conspicuous the trap the better. So far from being carefully concealed it should be placed in the open, though preferably near some bushes as the robin is shy of moving far from cover. For bait, stale white bread is as good as anything, because it shows up so well on dark ground. Sometimes the robin in whose territory the trap has been placed has been caught within two minutes of setting up the trap, and it was rare to have to wait more than two hours. This makes the robin ideally suited for an intimate study, for it took only a short time to trap every individual resident in the twenty acres of ground under investigation. No other British bird can be trapped so easily.

Robins can be caught in this way only in winter between about October and March, hard weather being a particularly suitable time. Winter trapping was relied on to get all the birds marked before the breeding season, and it was not necessary to trap any birds at the nest, which, though possible, might have caused disturbance or even desertion. To this there was one exception, a hen which arrived late in the spring and escaped trapping. When incubating, this bird was so tame that she allowed me to lift her off the eggs with my hand, put rings on her legs and replace her on the nest, where she settled down quietly to continue incubating. Such a degree of tameness is exceptional, but there are several similar cases on record.

Carefully carried out, trapping does not hurt the robins at all. The trap must be visited repeatedly so that the birds do not have to wait long inside, and each evening it must be closed or turned over so that birds do not get caught after the last evening visit and remain in all night. The last is a necessary precaution. One evening it was forgotten, and early next morning there was a tawny owl in the trap, evidently drawn there by two chaffinches, which must have got in in the late evening, and which the owl had killed and was eating.

The behaviour of the robins suggests how little they mind being trapped. If a trap was left for some days in the same territory the owning robin was usually caught at least once each day, and there was one bird which kept coming in. It was let out seven times one day and eight times the next. Towards the end, so soon as let out, it would sit about waiting for one to move off and then would promptly go in again. It proved such a nuisance that the trap, which was being kept there to catch other birds, had to be shifted. The reason for this behaviour is obscure, particularly as the robin did not feed when inside the trap, but simply perched on a bar and waited for its release. Although some robins were trapped over and over again, none ever learnt to find their way out.

When a robin was caught, two light, coloured, celluloid rings were placed on one leg. Rings were available in six plain colours and six striped in two colours. Using two rings on each bird, this provided seventy-eight combinations. After these had been exhausted, combinations of three rings were used. On the robin's other leg was placed one of the numbered metal rings formerly supplied by the 'British Birds' Marking Scheme and now by the Bird Ringing Committee of the British Trust for Ornithology. The purpose of the coloured rings was to enable me to tell which individual robin I was looking at without having to re-trap it, and of the metal one so that, if any of the birds should be found elsewhere, a report would be sent to the address on the ring. These rings are made specially light for the purpose, and observation showed that after an hour or two the birds apparently cease to notice them. Some robins carried the rings for three years and more, and seemed quite unhampered.

The above remarks refer to adult robins. Young birds in the nest were also ringed, but only with the metal rings, as so many of these juveniles die in the first two months of their life and many others move out of the area in which they are born. The few juveniles which remained in the following winter were trapped again then, and had coloured rings added.

Coloured rings provide much the most suitable means of marking robins individually. Through field-glasses or a telescope the different colours are clearly distinguishable at a distance sufficiently great for the observer to watch the birds without in the least disturbing them. In a behaviour study the latter is essential. If an observer wishes to watch a bird behaving naturally, he must keep much further away than is necessary when identification is his sole object.

Colour-ringing enables the observer to fit a particular piece of behaviour to the previous and subsequent behaviour of the individual bird concerned. It is particularly important in the robin as it provides the only means of distinguishing the two sexes. Cock and hen robin look alike, despite the popular belief, not extinct even now, that the robin's wife is the wren (the ox-eye, or great tit, was considered the wren's paramour). The statement in eighteenth-century books on birds that the hen robin shows less red or a duller red on the breast than the cock is also untrue. Nevertheless the sex of almost every robin watched during this investigation was certainly known; once ringed, each individual robin had a card to itself on which its behaviour was noted; in spring certain actions readily distinguish cock from hen (for instance, the cock feeds the hen and not vice versa), and, once having observed such diagnostic behaviour and noted the colour-combination of the rings of the birds concerned, the sex of each bird was established. The importance of this will be realized from the fact that several highly experienced field observers have described as courtship between the robin and its mate what is really the fighting of two rivals.

Incidentally colour-ringing also provides the bird watcher with a great deal of pleasure, and it is surprising that more amateur bird-watchers, especially those who put out food for the birds in winter, have not taken to it. For it enables the observer to know his birds individually in a way which is otherwise impossible. One suggestion I received was that, when they paired up, cock and hen robin might exchange rings.

During the four years of the investigation I ringed a hundred and nineteen adult robins which were each resident for at least a few weeks in the area of twenty acres to which the study was restricted. Of these birds one hen remained four years, several individuals for three years, quite a number two years, the great majority under a year. A few other adults which were not resident, but were either trespassing or wandering through, also got ringed. A hundred and twenty-one nestling robins were ringed.

None of these ringed robins was reported away from Dartington, and in general the proportion of ringed birds reported is so small that, if the study were being undertaken again, I would not use numbered metal rings at all. Instead I would put the same combination of coloured rings on both the robin's legs. A robin often perched so that only one of its legs was visible, and if this leg happened to be the one with the numbered metal ring, the bird's identity could not be determined until either robin or observer had changed positions. As a result, unnecessary time was sometimes wasted in establishing identity.

Certain colours proved much more suitable than others. Green and brown were difficult to see, and their use was abandoned. The most conspicuous colours were pink, white, and yellow, these same colours striped with black, and pink striped with white. Blue was fairly good, both plain and striped with black or white. Red was good but was abandoned as it was hard to distinguish from pink, and blue-and-white was given up because it could be confused with black and white. Clear view of a robin's leg was apt to be brief, and it paid to use strikingly different colour combinations for individual robins which were likely to come near each other, such as the cocks of adjoining territories or the two members of a pair. These coloured rings, made at the Greenrigg works, are available in two sizes, suitable for the smaller and larger garden birds respectively. After three years' exposure to the weather, they were almost as bright as when first put on. Fortunately the robin rarely gets them muddy.

Properly carried out, trapping and ringing do not harm the robins. It is in considerably greater fear of censure that I mention another method of investigation. 'Thou art a fool; the robin red breast and the nightingale never live long in cages,' Webster wrote. Nevertheless, after three years of study, I realized that certain aspects of behaviour would become much clearer if robins could be induced to nest in captivity. For this purpose two aviaries were constructed, each thirty feet long, one twelve, the other twenty feet wide, and each over six feet high. They had good grass turf, extensively planted with evergreen bushes, while branches were nailed to the walls higher up. Shelters were provided by wooden sheds and packing cases, half filled with brushwood. The birds were fed on Allen Silver's special insect mixture, together with large numbers of live mealworms, of which they are particularly fond. The food trays had wooden shelters over them to keep off the rain. Drinking-baths were readily provided by upturned dustbin lids. The chief danger in an aviary is from rats, and to guard against them the wire was sunk on every side to about a foot below the surface, and, when the earth was filled in, large quantities of broken glass were added. The rats did not get in, though extremely numerous round about.

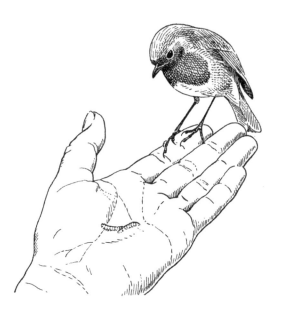

Two pairs of robins were placed in each aviary, in which lay the risk, for every experienced aviculturist who was consulted assured me that if two pairs of robins were placed together one pair was certain to kill the other. Indeed, one captive robin actually killed four others in four days. On the other hand, observations discussed later led me to expect that, provided the birds were given sufficient room in which to escape, one would not kill another, and that gradually their animosity would decline. The latter was what in fact happened and, while I was ready to remove one of the pairs at once if they showed any signs of being hurt, this was never necessary. Presumably other aviculturists have used much smaller aviaries.

The birds remained in good condition and in each aviary one of the pairs successfully reared five young, for which a prodigious number of mealworms was required. As soon as the young were well able to fend for themselves, they and their parents were let out. A year and a half later at least two of the adults were still alive in the wild state. I formerly thought that these were the first robins to breed in captivity in Britain, but there is one previous record, about seventy-five years ago.

The Life-History summarized

Adult robins moult in July and August, during which period they are retiring in their habits. Towards the end of July some of the young birds hatched in the previous spring begin to sing and to chase other robins, and the adults start to do the same about a fortnight later, after which each cock robin holds an exclusive patch of ground, the territory, in which it sings and from which it drives out all other robins. Some of the hen robins, like cocks, hold individual territories in autumn, while the rest migrate. Among the residents of both sexes, song and fighting continue throughout the autumn, gradually getting rather weaker. In late December and early January there is a marked revival of male song. Resident hens now leave their individual territories and there is an influx of other hens from outside. The pairs are formed between late December and early March, after which cock and hen share a territory.

About the middle of March the hen builds a nest, courtship follows, then egg-laying, incubation, and the feeding of the young. The pair rear a succession of broods until June, after which they retire in preparation for the moult.

The dates in this summary refer to South Devon close to sea-level. Events occur rather earlier here than in other parts of England, and they occur later the farther north one goes. In addition to this, there are small differences in the same area from one year to the next, and considerably greater differences among individual robins. Thus some juvenile robins begin their autumn song at least a month before others, and smaller variations affect all the other times recorded.

The main outlines of the life history of the robin were elucidated by J. P. Burkitt at Enniskillen in Ireland, some ten years before the present study was commenced. Burkitt was the first observer of birds to use coloured rings, and must be given full credit for his valuable and original work. His papers, together with my own on the robin, are listed at the beginning of the references (see p. 249).

The succession of events may be amplified by describing the happenings throughout the year in one particular isolated copse. After the robins were ringed in early October this copse was found, as shown in Fig. 1 (overleaf), to be divided among five robins, which for convenience will be referred to from west to east along the copse as A, B, C, D, and E. At the beginning nothing was known of any of these birds, for all happened to be new arrivals since the last breeding season. (This is rather unusual.) A, C, D, and E sang well; B was never heard to sing.

On 20 December C suddenly came into loud song, drove out D from his territory and at the same time paired up with a new arrival, which will be called F. C (a cock) and F (a hen)

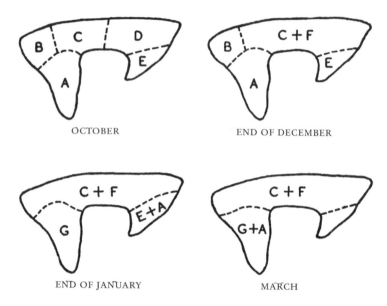

OCTOBER

END OF DECEMBER

END OF JANUARY

MARCH

Fig. 1. Changes in Ownership

now shared the territory which included C's old territory together with that formerly occupied by D (presumably but not certainly a cock), who was never seen again.

In the second half of January further changes took place. A, who had been in good song throughout October, now revealed herself as a hen, for she left her territory and moved to the opposite end of the copse to form a pair with E (a cock). Soon afterwards B, who was not heard to sing in October, also left her territory and paired up with a cock across the meadow in the next copse. The departures of hens A and B left their former territories vacant, and this ground was claimed a few days later by a new arrival, G (a cock). Soon afterwards cock E was killed by a cat, and his former mate, hen A, now moved back to her old territory and paired up with its new occupant, cock G. Cock C, with his mate, hen F, expanded into the extra territory left vacant by A's departure. For the breeding season, therefore, the copse was now occupied by two pairs, cock G with hen A on the west side, and cock C with hen F on the east side. Similar changes occur everywhere, though not always so complicated.

ROBIN SONG

Autumn

And now with treble soft
The redbreast whistles from a garden croft.

JOHN KEATS

Mid-Winter

The redbreast warbles still but is content
With slender notes and more than half suppressed:
Pleased with his solitude, and flitting light,
From spray to spray, where'er he rests he shakes
From many a twig the pendent drops of ice,
That tinkle in the withered leaves below,
Stillness accompanied with sound so soft
Charms more than silence.

WILLIAM COWPER

Early Spring

Robin on a leafless bough,
Lord in Heaven, how he sings.

W. H. DAVIES

Summer

The merry Larke hir mattins sings aloft,
The Thrush replyes, the Mauis descant playes,
The Ouzell shrills, the Ruddock warbles soft,
So goodly all agree with sweet consent,
To this dayes merriment.

EDMUND SPENSER

2

Song

Singing is a good and proper thing. You may be sure
your father would not sing if it were not so.

MRS SARAH TRIMMER: *A History of the Robins*
(1786)

Among song-birds classical tradition assigns first place to the
nightingale. As Pliny wrote, 'There is not a pipe or instrument
againe in the world (devised with all the art and cunning of
man so exquisitely as possibly might be) that can affoord
more musick than this pretty bird doth out of that little throat
of hers.' But many would agree with Nicholas Cox, who said
of the robin redbreast, 'It is the opinion of some, that this little
King of Birds for sweetness of Note comes not much short
of the Nightingale.' Indeed, through its homely associations
and its habit of singing almost throughout the year, the robin
might well come first in popular opinion today.

The spring song of the robin starts near the end of December and continues until about the middle of June. The autumn song, which is thinner and less rich, is first heard in late July from some of the young birds, the adult robins starting about a fortnight later. It is continued throughout the autumn, but is rather feeble in the early winter, until the spring song suddenly starts again in late December. Thus robins sing throughout the year except for a gap between mid-June and mid-July, but an occasional late adult or early youngster can sometimes be heard even in the latter period, and once on 13 July I heard an early juvenile singing the autumn song against a late adult singing the spring song. The above dates refer to South Devon and are rather different in other parts of England.

The song of the robin is one of the most characteristic sounds of the English countryside in autumn, and the robin is the only British bird to sing persistently then. At this season not only the cocks but also about half of the hen robins sing, their song being indistinguishable from that of the cocks. Other hens sing poorly and some are not heard at all. After a hen has paired with a cock in spring it very rarely sings.

The robin's song shows considerable individual variations. There are good and poor singers. Some start earlier in the season than others, or continue later. In spring unmated cocks usually sing better than mated ones, but a few mated males have sung very well. Robins have occasionally been heard to sing at night.

The robin's spring song can be heard on a gramophone record made by Ludwig Koch. Analysis of the sound-tracks of bird song has shown that most small birds include in their songs notes of a frequency well above the limits to which the human ear is sensitive, so that a bird's song probably does not sound the same to another bird as it does to us. Nor does it sound the same to every ornithologist, since the sensitivity of the human ear to these frequencies varies markedly with the individual, and declines with increasing age.

Why do birds sing? The most popular answer is because they are happy. From which it could be concluded that whereas cock robins are happy most of the year the hens are happy only in autumn, that cock robins are happier before than after obtaining mates, and that they are happiest of all when fighting. 'Premises assumed without evidence or in spite of it; and conclusions drawn from them so logically, that they must necessarily be false,' as Peacock puts it. A bird's happiness is unobservable and the question 'Is a bird happy?' is impossible to answer and perhaps meaningless. The other popular view, that the cock bird sings in courtship to please the hen, was finally disproved by Eliot Howard, who showed that the chief function of bird song is to advertise the cock in his territory, both to other cocks which may intrude and to hens in search of mates.

Observation soon establishes that each colour-ringed male robin sings only in its own territory, and by noting on a map every place in which each bird is seen singing, the boundaries of the different territories are quickly determined. It is impossible to drive a singing robin from its territory. As the observer approaches the bird retreats, but on reaching the edge of its territory it does not proceed further, and if chivied it unexpectedly flies back over the observer's head to the middle of its ground.

The aviary experiments showed the connexion between ownership of territory and song, since in each aviary there were two pairs of robins, and in each case one of the two cocks became the owner of the whole aviary and it was this bird which did nearly all the singing. In each aviary for a short time near the beginning of the experiment the second male temporarily became the owner, and then it was this second male which sang and the other was silent. While ownership was in dispute, both cocks sang.

The occurrence of autumn song also shows the connexion between song and territory, for at this season courtship does not occur, but each cock robin holds a territory, while those hens which sing in autumn are, like cocks, defending individual territories. Only two hen robins were heard to sing strongly in spring after pairing up, in one case when a pair moved into and claimed a new territory (Burkitt recorded a similar incident), and in the other case when a pair were defending their territory against a rival pair. The correlation between song and territory is again clear.

The fact that hen robins sing has found its way into ornithological works only since Burkitt proved it in 1924 with his colour-ringed birds in Ireland, though actually this possibility was mentioned in 1831, and was stated more definitely by Charles Darwin in 1871. Female song is uncommon in birds. Among European species it has been recorded in the starling, as discussed in Chapter 11, and in the dipper. Perhaps some of the autumn song of skylarks comes from hen birds, since these, like the female robin, have been recorded as singing in captivity. An occasional hen blackbird or chaffinch also sings in a wild state, but only as an abnormality, and the females of a few other species have been heard in captivity. Possibly future observations will show that there are other British birds in which the female sings, for there are many birds in which, like the robin, the sexes cannot be distinguished by plumage, and most observers assume that a singing bird is necessarily a male. Female song also occurs in various

North American species. The hen mocking-bird sings regularly in the fall, as does the loggerhead shrike. The mocking-bird is one of America's most famous song-birds, and at the present time the song is frequently heard in England, as it is the thrush-like song regularly introduced into Hollywood films for sentimental occasions. In both mocking-bird and loggerhead shrike the hen holds an individual autumn territory, thus providing further confirmation of the connexion between song and territory.

The most important use of song to the robin in its territory is to advertise possession to rivals and to warn them off. When an intruding robin comes close to the boundary or actually trespasses, the owner's song becomes specially loud, the intruder often retreats at once, and in this way song saves many fights. To cite an example: On 27 May 1937, an un-ringed newcomer robin, evidently wandering without territory, started to sing in a corner of the territory owned by a long-established resident male. The latter, then in a distant part of its territory, promptly sang in reply. The newcomer, which could not, of course, yet know that it was trespassing, sang again. The owner, having flown rather closer in the interval, sang again in reply. The newcomer again sang, the owner again approached and replied, now more vigorously, and this procedure was repeated twice more, the owner finally uttering a violent song-phrase from only some fifteen yards away, but still hidden from sight by thick bushes. At this point the newcomer fled, from an opponent it never saw, nor did it appear again.

With the above incident, compare the following from H. G. Wells' *Outline of History*: 'It was three o'clock on 14 August 1431, that the crusaders, who were encamped in the plain between Domazlice and Horsuv Tyn, received the news that the

Hussites, under the leadership of Prokop the Great, were approaching. Though the Bohemians were still four miles off, the rattle of their war wagons and the song "All ye warriors of God" which their whole host was chanting, could already be heard. The enthusiasm of the crusaders evaporated with astounding rapidity. Lützow describes how the papal representative and the Duke of Saxony ascended a convenient hill to inspect the battlefield. It was, they discovered, not going to be a battlefield. The German camp was in utter confusion. Horsemen were streaming off in every direction, and the clatter of empty wagons being driven off almost drowned the sound of that terrible singing… So ended the Bohemian crusade.'

The influence of song in the fighting of the robin was also shown in connexion with the aviaries. When captive robins were put in one of the aviaries, the wild owners of the ground outside perched on the aviary roof and sang and displayed at the birds inside. After a few days one of the captive male robins came into full song, singing particularly whenever the wild male approached the aviary, and after a few days the latter retired, leaving the captive male in undisputed possession not only of the aviary but of the ground immediately around it. A similar series of events occurred at the other aviary. This result can be attributed only in part to the singing of the captive robins, since a wild robin gradually ceases to attack an intruding robin which will not retreat out of its territory, a point discussed in Chapters 12 and 13. But song can be given full credit for subsequent victories. During April the territory in the copse outside one of the aviaries fell vacant. A cock robin soon arrived, claiming the vacant ground with loud song. The aviary male, which had been rather quiet for some time, promptly came into full song, and so vigorously that the newcomer made no serious attempt to acquire the ground immediately round the aviary. A similar incident occurred outside the other aviary a month later. Hence through singing the aviary males were victorious against wild rivals which they could never attack directly, and conquered territory on which they could never settle.

Not only does the song of the robin serve as a warning prelude to a fight, but robins actually sing while fighting, interpolating vigorous song-phrases between their attacks on intruders, while the finest singing of the year is heard when one cock is trying to establish itself in the territory of another. At least in the robin, the chorus of song early on a spring morning – often called the birds' Hymn to the Dawn – is a hymn of battle rather than of love.

Similarly war-whoops, gongs, trumpets, and the pibroch have played no inconsiderable part in human battles. Marco Polo's description of one of Kublai Khan's engagements is typical: 'As soon as the order of battle was arranged, an infinite number of wind instruments of various kinds were sounded, and these were succeeded by songs, according to the custom of the Tartars before they engage in fight, which commences upon the signal given by the cymbals and drums, and there was such a beating of cymbals and drums, and such singing, that it was wonderful to hear . . . and then a fierce and bloody conflict began.' The value of such methods is demonstrated by the account of the Bohemian crusade already given. Curiously, this function of robin song was mentioned in a verse published anonymously in 1834, but since overlooked:

> O blind to Nature's all accordant plan,
> Think not the war-song is confin'd to man;
> In shrill defiance, ere they join the fray,
> Robin to robin chaunts the martial lay.

Warning to rivals is not the sole use of song in the robin, though it is the most important. In the early spring the song also serves to advertise the unmated cock in possession of a territory to hens in search of mates. 'Ful loude he sang – Com hider, love, to me' – like the Pardoner of Rouncivale. Song is the chief way in which the hen robin can locate an unmated cock, and almost immediately after the cock has obtained a mate its song declines to a rather moderate intensity and remains so, except during fights, unless the mate is lost, in which case the cock again comes into loud song. In the late spring nearly all the robin song comes from cocks which are unmated.

In some other species of birds the song of the male is loud until pair-formation, and thereafter becomes extremely poor. Thus Mrs Nice writes of the American song-sparrow: 'I often say to myself on nearing a territory where silence reigns over night, "Such and such a male must be either dead or married," and upon careful search I find either two birds or none.' Evidently in such species advertisement for a female is a much more important function of song than is the warning off of a rival male from the territory.

Colonel Montagu (1802) of whom Charles Darwin wrote, 'Few more careful observers ever lived', was the first to realize this function of bird song. 'The males of songbirds, and many others, do not in general search for the female, but, on the contrary, their business in the spring is to perch on some conspicuous spot, breathing out their full and amorous notes, which, by instinct, the female knows, and repairs to the spot to choose her mate.' Montagu observed the marked decline in song after the cock acquires a mate and showed experimentally in the redstart that if the mate was removed the cock returned to full song. These facts were quoted by Darwin, but since then appear to have been overlooked until rediscovered by Howard in the twentieth century.

It need hardly be stressed that this second function of song is to advertise for, and not to please, a hen. So soon as a hen is obtained the song declines, and in most birds it plays no part in the subsequent courtship. There are, however, a few birds, but not including the robin, in which the cock sings while courting the hen. I have observed this in one of the American goldfinches, and it is true of other finches in the linnet group. In some of these finches, also in barbets, American wrens, and others, the hen may sing antiphonally with the cock, another example of female song in birds.

Once when I was trying to catch an elusive robin in the house-trap the bird burst into song as it ran about the ground, and it continued to sing for a little after I had caught it and was holding it on its back in my hand. It is well known that other birds will occasionally sing or display when alarmed. A sudden thunder-clap or bomb often starts them off. As a more spectacular example, when in an Imperial Airways machine over the Kenya Game Reserve, we on several occasions flew close to a male ostrich, at which the latter would go down in the sand, spread its white plumes, and rock gently from side to side in display at the aircraft.

Occasionally robins will sing in answer to the song of other species of birds, but usually only when no other robins are singing. One hen robin in autumn used to get particularly excited by the singing of a tame canary in a room bordering her territory, and she would come down to the cage and sing in answer. More curious, a correspondent wrote to me of a robin which used to sing whenever he used a double-handed cross-cut saw. The latter has a characteristic note, and the robin regularly sang back at it, stopping so soon as the sawing stopped.

A puzzling feature of robin singing is the subdued song sometimes heard in autumn. A robin may utter quiet phrases, audible at only a few yards, continuously for several minutes. Several other species sometimes sing in this apparently purposeless way in autumn.

The song of each species of bird is distinctive from that of every other found in the same region with it. This arrangement, so convenient for the ornithologist, is no accident. To quote Montagu again: 'The peculiar note of each is an unerring mark for each to discover its own species.' It is essential that the female should recognize a male of its own species. Such recognition has survival value since hybrids between species are rarely fertile and may also be less efficient in other ways.

Of course there is no reason for a bird's song to be distinct from that of a species with which it never comes into contact. The song of the Galapagos mangrove warbler was continually reminding me of that of a robin, and one of the East African sunbirds sings very like a willow warbler. Bird song has considerable subjective associations, and it was curious in the heat and dust of the tropics to be suddenly transported back to a Devon spring.

A few of the South Devon robins, though not many, incorporated the notes of other species into their songs. Thus one used a phrase reminiscent of the song of the chaffinch, and another of the spring 'teacher' call of the great tit. Witchell has listed the songs and calls of other species of birds which he states that he has heard in different robins' songs. More extraordinary is the report by John Morton in 1712. 'Besides the common Sort of Singing Birds... the Ingenious

Mr Mansel has had… a Robinredbreast that not only learnt some Flagelet Tunes, but spoke distinctly several short Sentences.' This account I at first rejected as incredible, but since then have read Gesner who quotes Porphyry (*De Abstinentia esu Animalium*, *lib. 3*, third century AD) to the effect that crows, magpies, and robins imitate man and remember what they hear; and in 1823 there was a lady in Edinburgh whose tame robin very distinctly pronounced 'How do ye do?' and several other words. Other song-birds can also be trained to speak. Pliny noted that 'Agrippina the Empresse, wife to Claudius Cæsar, had a Blackbird or a Throstle, at what time I compiled this book, which could counterfeit man's speech. The two Cæsars also, the yong Princes (to wit Germanicus and Drusus) had one Stare and sundry Nightingales taught to parle Greek and Latin.' Barrington knew of a linnet which sang 'pretty boy'.

Such occurrences raise the question of whether bird song is inherited or learnt. Aristotle already knew the answer. 'Of little birds, some give a different note from the parent birds, if they have been removed from the nest and have heard other birds singing.' It is rather surprising that almost the only experiments on this subject in British birds were made in the late eighteenth century by Daines Barrington, best known as one of Gilbert White's correspondents. Of the robin Barrington wrote: 'I educated a young robin under a very fine nightingale; which, however, began already to be out of song, and was perfectly mute in less than a fortnight. This robin afterwards sung three parts in four nightingale; and the rest of his song was what the bird-catchers call rubbish, or no particular note whatsoever.' Barrington got another captive robin to sing like a skylark-linnet', *i.e.* like a linnet which has been raised under a skylark, and so sang like the latter. His experiments show with admirable clearness that in the species with which he worked the specific characteristics of the song are learnt by the young bird from the birds which it bears during and shortly after fledging. This suggests that those wild robins which imitate the songs of other species may have heard the latter along with the song of their parent when they were young birds.

Experiments in the United States fully confirm Barrington's results for a large number of other song-birds, and show also that, when such birds are raised without hearing any other birds sing, they sing an original type of song quite unrecognizable as that of their own kind, *i.e.* 'what the bird-catchers call rubbish.' Even the chirping house-sparrow can be made to sing – how excellent if Britain's cities could be filled with 'nightingale-sparrows'.

Observations in Germany show that there are other species in which the song cannot be changed. Examples are the swallow, chiffchaff, and grasshopper warbler, and the same must apply to the cuckoo, since this bird is not raised by its own kind. Lloyd Morgan stated that the song was also inherited in the song-thrush, but later work has shown that in this species the song can be slightly modified. It also seems probable that, even in those species in which the song is highly modifiable, there is an inherited predisposition to learn the correct song. For instance, a nightingale first sang like the other species with which it had been reared, but rapidly changed to the normal song when it heard it in the following year.

For long I was puzzled to account for the sudden marked increase in the cock robin's song just after the young leave the nest and while they are still under his care. This happened both in the wild and in the aviaries. I suggest that song at this period may have survival value because it is at this stage that the young robins learn the song of their species, though they do not themselves sing then. The young robins start to sing in late July, before the adults, so they get no other chance to learn their song. Observations by Burkitt and others show that in other song-birds there is a similar revival of song about the time that the young leave the nest, and perhaps it has the same function as that suggested here for the robin. In all those species in which the song is not inherited but is acquired by the young bird, it is clear that the young must get such an opportunity as this to hear the song before they leave their parents' charge.

3

The Red Breast

Fat and merry, lean and sad,
Pale and pettish, red and bad.
Proverbs of Alfred

Its orange-red breast is the robin's most conspicuous feature,
and indeed was responsible for its earlier name. Through its
friendly familiarity the redbreast became robin-redbreast,
and there is now an increasing tendency to drop the redbreast
altogether. In the same way the pie was called Maggie, but in
this case both words have been retained in the modern name.
The Saxon 'ruddock' also refers to the robin's red colouring,
but this and the sixteenth-century 'robinet' have disappeared
from general use.

Young robins are speckled brown when they leave the
nest and the red breast is acquired by a moult in late July or
August, after which all robins of both sexes have a red breast.

Below, the feathers of the abdomen are white, and sometimes one of them obtrudes on to the breast, like

> An erring Lace, which here and there
> Enthralls the Crimson Stomacher.
>
> *R. Herrick*

The robin's breast does not become a fuller crimson in spring, despite the statement so often quoted from *Locksley Hall*, and, in view of the real function of the red breast, this was an unfortunate parallel with which to open a love poem. It may also be mentioned that the 'red red Robin' which goes 'bob-bob-bobbin' along' is, of course, not the British, but the American robin, a thrush misnamed by homesick emigrants on account of its brick-red underparts. There is hardly a corner of the world in which the English have not managed to find some redbreasted bird which they could call a robin.

When a bird possesses a bright patch of colour, one may guess that it plays a part in its life sufficiently important to outweigh the disadvantage of added conspicuousness to enemies. The red breast of the robin is used in display.

The display of the robin is usually described as rare, and few ornithologists have seen it. Actually it is exceedingly common, especially in the early morning. To see it early rising is not essential, but birds live much more intensely around dawn than at any other time of day. This is fortunate, as it means that the study of birds can be combined with a normal working day. Dawn watching has many pleasures as so much life carries on undisturbed around one. Rabbits have jumped over my feet, and once, only ten feet away, I saw a young rabbit taken by a weasel. Concealment is normally unnecessary but stillness is essential, and should one wish to turn in another direction or to raise the field-glasses or telescope to the eye, it must be done gradually, however urgent the occasion. Sudden movement, while it may not cause the birds to depart, usually makes them too nervous to behave naturally.

The robin's display has usually been described as courtship, and as such it has been illustrated in at least two books on birds and in the journal of the British Ornithologists' Union. To an earlier generation, love was apparently the only emotion which could look absurd. Colour-ringing soon showed that this display was not courtship at all, since the two individuals between which it almost never occurred were the members of a pair. On the other hand, it was regularly given by the owner of a territory of either sex towards an intruding robin of either sex. It is, in fact, a threat display, serving to intimidate a trespassing robin and to drive it away. Like song, but coming into operation at close quarters, its function is to save actual fighting. From his study of ringed robins in Ireland Burkitt earlier came to the same conclusion, but from casual observation several later writers have repeated, quite erroneously, that it is a courtship display.

When the owning robin sees an intruder in its territory it promptly flies at it, sometimes uttering a preliminary loud song-phrase. If, as often happens, the intruder flies off at once, the owner pursues it out of the territory. If, however, the intruder does not retreat, the owner flies right on up to it, but usually, instead of striking, stops a foot or two away and displays. Only if this display fails to make the intruder depart, and it rarely fails to achieve this, does the owning robin come to blows with the intruder.

The display of the robin is remarkable. Much the most prominent feature is the red breast, which is stretched and so held that the intruding robin sees as much of it as possible. If the attacking robin is above the intruder, it thrusts its head forward so that the red breast is directed downward at the intruder. On the other hand, when the aggressive owner happens to perch below the intruder, head and neck are stretched upward so that the beak points skyward or even backward, as a result of which the intruder is again presented with the maximum area of red. When the owner is at the same level as the intruder, its attitude is between these two extremes. The main display attitudes are shown in Fig. 2 overleaf, based on photographs taken for me at Dartington by Mr H. N. Southern, which were the first photographs of the robin's display to be published. We utilized the fact that, as described in Chapter 12, a wild robin will display at a stuffed robin placed in its territory.

On one occasion the only perch left close to a stuffed robin necessitated the attacking owner perching sideways on. But even here it achieved the attitude which presented the intruder robin with the maximum area of red breast, by skewing the stretched neck round into a most awkward position. A displaying bird usually assumes the attitude which exhibits its colours to the best advantage, but the robin seems unusually adaptable in this respect (see also the chapter head on p. 36).

Still photographs do not give an adequate picture of the robin's display, since movement is an essential part of it. Keeping its feet still the posturing robin may slowly turn its body, and in particular the red breast, from side to side through an arc of nearly a hundred and eighty degrees. This slow and steady movement is particularly striking in a bird that normally moves rapidly and jerkily. When head and tail are both vertical the effect is fantastic. There are also variations. Sometimes the bird turns through a smaller arc more

rapidly. Sometimes, again, instead of turning from side to side, it jerks its head and neck vertically up and down, and this may be combined with a rapid movement of the feet, the bird appearing to dance around the intruder. Other performers are much quieter, and some robins merely keep still with a feebly stretched neck, so that it takes an experienced observer to realize that they are displaying at all.

The red breast is the most prominent but not the only part of the body involved in the display. The wings are sometimes flicked half open, the tail may be cocked, the feathers of the crown erected. But these are subsidiary movements which are often omitted, and the feathers involved are dull-coloured except for a small orange spot on the angle of the wing. To some extent all the feathers are fluffed out, making the attacking bird appear appreciably larger than usual, and the display is often accentuated by song-phrases, which may be normal in tone but more often take the form of a characteristic high pitched squealing.

The hen robin, like the cock, has a red breast. It is uncommon among birds for the female to be as brightly coloured as the male. It is also uncommon for a female to fight as much as the hen robin, who both helps to drive intruders from the shared spring territory and owns an independent territory in autumn. When fighting, a hen robin displays with her breast in just the same way as a cock.

The nest of a robin is usually concealed in a hollow, so that when the hen is incubating the red breast is hidden and does not draw attention to the nest. A brightly coloured female would be much more dangerous for a species which

Opposite: Fig. 2. Threat display. Note difference in attitude according to whether the attacking bird (on left) is above, level with, or below the intruder

nested in the open. Similarly, the cock robin makes its red breast as inconspicuous as possible when danger threatens. I have twice seen a singing robin stop suddenly, and flatten itself out along the branch on which it was perched. On one of these occasions I looked up to see a kestrel hovering overhead. The robin's attitude would have made it less conspicuous from above.

The threat display of the robin is certainly effective. The intruding robin usually departs almost at once, and rarely does the attacking bird have to change from posturing to direct attack to achieve this. Exhibition of the red breast suffices. With which, compare the following, related in *Tibetan Trek* by Ronald Kaulback. The coolies employed on this expedition came in to the leader, Kingdon Ward, breathing fire and murder, because they considered their pay and rice rations insufficient. 'Kingdon Ward had a magnificent scarlet sweater of generous proportions, which he had put on to baffle the draughts. Seeing him so gorgeously arrayed, about three-quarters of the malcontents' bluster instantly died away, and the argument, though prolonged, was conducted quite peaceably from then on.'

The robin's red breast, which is associated in legend both with fire-bringing and with the Cross, really plays an important, though rather different, part in the life of the bird. Just as its song is a war cry, so its red breast is war paint, both song and plumage helping to prevent a fight coming to blows. The red uniform of the British soldier probably served a similar purpose, though also, of course, a recognition mark.

Darwin's essay on sexual selection was the first serious attempt to interpret the bright colours and complex ornaments of male birds. Subsequently naturalists tended to assume that all displays and bright colouring were associated with courtship. Many are, but, as Hingston pointed out, though he goes too far in the opposite direction, many others are threat display. For example, the cock chaffinch sways towards a rival male, partially opening the wings out sideways, which exposes the white patch on the shoulder. The goldcrest erects and spreads its orange-red crest so that the head seems on fire. The great crested grebe stretches head and neck horizontally along the surface of the water and spreads its chestnut ruff. The performance of the male domestic turkey is too well known to need description. In some species, like the robin, the bright coloration functions only in threat display. In others, such as the great crested grebe and the blackcap, the conspicuous colours are used both in threat and courtship. In the great crested grebe, though the same coloured areas are displayed, the threat and courtship attitudes are quite different, but there are other birds in which they are said to look alike.

The threat and courtship displays of birds include various types of attitudes and actions. These probably owe their origin to different causes. First, there are generally excited movements which may involve almost any part of the body. If violent action such as that of fighting is deferred, this might be expected to result in random body movements and the erection of feathers generally, as occurs in threat display.

Secondly, habits are incorporated which are either meaningless or are really appropriate to quite different behaviour. For example, in the threat display of the herring-gull the two rivals bend down and tear out large quantities of grass. Preening, the picking of buds, feeding, drinking, and toying with nest material are other inappropriate activities which sometimes feature in threat or courtship display, in which they probably owe their origin to the fact, mentioned in the last chapter, that when extremely excited a bird sometimes performs some quite meaningless action. Tearing the hair in distress or scratching the head in bewilderment are human equivalents.

Thirdly, some of the attitudes in display are suggestive of what the bird is about to do. In the threat display of many species the head is lowered and extended, as it is in a bird about to launch a real attack, or, as in the chaffinch, the wings may be partly opened as if it were about to fly at the rival. That birds appreciate the significance of preparatory attitudes was stressed by O. Heinroth, the German ornithologist, who describes how, when one member of a flock of geese or ducks adopts the attitude preparatory to flight, the others are quick to appreciate it and follow suit. Similarly an intruding bird will appreciate the significance of the owner's attitude preparatory to attack and may anticipate the latter by flight. Hence it is readily seen how a preparatory attitude may come to form part of a threat display.

While display movements have probably originated in these three ways, they have subsequently become of value to the species as a means of communication, as signals for particular situations, and many of the actions have become formalized and symbolic. Thus a displaying bird may perform actions which are recognizably those of preening but without actually preening its feathers. Again, the preparatory attitudes often seen in display do not, in fact, usually lead on to the actions which they suggest, and further they are often an exaggeration or intensification of the normal preparatory attitudes. The same is seen when two small boys prepare to engage in doubtful combat. Again the grass-plucking of the herring-gull may originally have been meaningless, but it is far too regular a part of the fighting to be meaningless at the present time. Its function, like the breast-waving of the robin, is as a threat display, to save a fight coming to actual blows.

Display attitudes and movements are made more conspicuous by brightly coloured feathers, and there is a tendency for the latter to be evolved on any part of the body involved in display. Once such brightly coloured areas have appeared there may be survival value in further elaborating the movements which display them, and thereafter colour and movement will tend to evolve together. The breast-waving of the robin and the erection of the tail coverts by the peacock have probably attained their present form in this way, as they are striking and quite peculiar movements which show off to the full the brightly coloured feathers. Very possibly they originated in one of the three ways discussed in preceding paragraphs, but if so they have now become so altered that their origin is unrecognizable.

The bright colours displayed by male birds are usually distinctive for each species. The chief reason seems the same as in the case of song, that these bright colour patterns serve as recognition marks, particularly for the females in search of mates. A cock which acquires a hen of the wrong species will rarely, if ever, produce fertile offspring. The bright colours are used in display, but there is the further need for the display of each species to be distinctive from that of others. That a colour pattern, when present, is the chief way in which a bird recognizes a member of its species is demonstrated in the case of the red breast of the robin in Chapters 12 and 13. In this connexion it would be most interesting to know what happens to those freak robins which are occasionally seen either lacking red on the breast or white all over. Unfortunately such birds have nearly always been shot at sight by collectors.

Display movements in birds, like language in mankind,

are a means of communication, and they are therefore 'conventional'. It does not greatly matter what form they take provided it is characteristic of one particular situation. Hence just as the same sound can carry a different meaning in different human languages, so a display movement may refer to a different situation in different species of birds. The same idea is illustrated by the very different meaning attached to the sideways movement of the tail in a dog as compared with a cat. And just as it is possible to trace the changes in a particular word through related human languages, so Heinroth and Lorenz have traced changes in particular display movements in different species of ducks, thus throwing further light on their relationships.

Further, just as some words have not only a descriptive but also an emotional content, so display is more than just a signal. For if the bird to which the display is made does not respond appropriately, the first bird may repeat its signal, often in an exaggerated form, and this sometimes achieves the appropriate effect when the first attempt failed. Thus the most violent robin postures occur when the intruder does not immediately retreat, and the most elaborate courtship displays occur in one bird when the other will not adopt the correct attitude for coition. Just how repetition or elaboration of a display causes another bird to do something of which it previously seemed incapable is far from clear. The display attitude might be said to 'stimulate' the other bird, but this term is merely a cloak for ignorance. One is here up against one of the basic problems of behaviour.

4

Fighting

It is a great victory that comes without blood.
Outlandish Proverbs, coll. by G. HERBERT (1639)

The male robin Black-and-white Pink-and-black (which will
be referred to as the first male) happened one day to enter a
trap in its own territory. A little later Double Blue, the cock
owning the next-door territory, entered the first male's territory on a trespassing excursion for food and came down to
the trap containing the first male. The latter at once postured
violently and uttered a vigorous song-phrase from inside the
trap, and Double Blue, who was one of the fiercest of all the
robins, promptly retreated to his own territory, although the
first male could not, of course, get out to attack him. I now
caught the first male and moved him into a trap in the territory of Double Blue. The formerly timid Double Blue now
came raging over the trap, posturing violently at the first male

inside, while the latter, formerly so fierce, made himself as scarce as possible, and did not attempt to fight or posture back. In both encounters the same two individuals were involved, so this experiment clearly demonstrates the importance of territory in the fighting of the robin.

Much the most frequent type of fighting among robins is this territorial attack by the owner of the territory on intruding individuals, and normally, as in the above instance, the owner of the territory is successful, and the intruder puts up no resistance. Victory, in fact, goes not to the strong but to the righteous, the righteous, of course, being the owner of property.

Considering that robins defend their territories so closely, the extent to which they trespass is remarkable. Robins repeatedly intrude into neighbouring territories in search of food, though promptly attacked if seen by the owner. One bird was trapped as much as three hundred yards from its own territory. I have also seen a trespassing robin vigorously driven away by the owning male from a particularly favourable feeding area crowded with insects, but only ten minutes later the same intruder was back again, keeping mainly in the cover and acting rather nervously, but feeding on the insects all the same. The owner of the territory was now elsewhere, so on this occasion the trespasser got away with it.

Since robins trespass so frequently, it might be asked how one can know that they own territories at all. The answer lies in the birds' behaviour. Inside their territories the birds sing, fight, display, and make themselves conspicuous; outside them they do not sing or display, they retreat if attacked, keep as inconspicuous as possible, and, if disturbed, usually fly straight back to their own domains.

Territorial attacks are specially frequent when a robin has newly arrived. For instance, a hen which has just paired up with a cock in early spring is at first unaware of the boundaries of her mate's territory. She wanders in all directions, and whenever she trespasses is promptly threatened by the neighbouring owner. Hens soon come to recognize places where they are attacked. Thus one hen was vigorously driven from a tree which happened to be just inside the neighbour's boundary. Twice in the next ten minutes she again began to fly towards the same tree, but then sheered off in mid-air as if recollecting her former reception. Newly arrived birds take only a day or two to learn all the boundaries of their territories, but before they come to do so they provide some good fighting, especially when, as sometimes happens, the newly mated cock follows his hen over the territory boundary. Four birds may then be involved in a scrap.

Boundary disputes are regular even between well established neighbours. If one singing male approaches the boundary of his territory the cock of the adjoining territory almost always approaches too, and the two birds sing loudly and posture vigorously at each other. After a while they retire again, and it is extremely uncommon for a boundary dispute to lead to actual blows, and then they are seldom serious. Nowadays it is usually assumed that intelligence is greatly superior to instinct; but while in the robin, with its instinctive behaviour, a complex type of bluff fighting has been evolved, it is rare indeed in intelligent man that a territorial dispute is settled bloodlessly. As William Cowper noted, 'Reasoning at every step he treads, man yet mistakes his way; While meaner things whom instinct leads are rarely known to stray.'

During the moult, adult robins show hardly any aggressive behaviour, though a little ineffective chasing is sometimes seen. In late July, when the juveniles acquire the red breast and start to sing, they also start to fight. Aggressive behaviour is at first sporadic. A nearby robin is chased one moment and ignored the next. Early in August some of the adult robins also begin to sing and chase other robins. They look rather untidy as they are still moulting heavily, but not so odd as the juveniles, which show an irregular patch of red on otherwise speckled brown plumage. After a week or two the birds become more aggressive; each limits its attacks to a particular area from which all other robins are driven, and from then on the birds may be said to own definite territories. At this season, as already noted, some of the hens are quite as fierce as cocks, but others, though keeping to restricted territories, are rarely seen to fight. Territories are maintained throughout the autumn and early winter. There is much fighting in the early autumn, especially by the juveniles, which employ more direct fighting and less posturing than the adults, and fierce encounters occur. By October things have greatly quieted down, partly owing to a decline in the fierceness of the robins, partly because the territorial boundaries are now settled and the birds, though they still trespass frequently, do so with more circumspection.

After mid-December the resident female robins leave their individual territories and pair up with males, and from then on the pair maintain a joint territory. The unmated cock is rather mild, but after pair-formation his fierceness increases considerably. During January and February most mated females are rather mild and take little part in driving intruders from the territory, though a few are fierce. But in the middle of March when nest-building commences the hens come out of retirement, and many are then as fierce in defence of the territory as cocks, though some remain mild. During incubation the hen rarely attacks intruders, but she again becomes aggressive when there are young in the nest.

In the robin both cock and hen attack intruders of either sex; they do not confine their attacks to members of their own sex. This is not the case in all territorial birds. Thus the cock chaffinch or snow bunting confines its attacks to trespassing cocks, the hen to trespassing hens. However, these species differ from the robin in that the two sexes are readily distinguishable by plumage differences. There are other territorial species in which only the male fights, while the hen takes no part at all and may, indeed, remain unaware of the territorial boundaries.

The aggressiveness of the robin begins to decline towards the end of May, and in June it is sometimes possible to see two pairs collecting food for their young near to each other without exciting any friction through trespass of territory. But a few individuals remain fierce. For example, in the last week of June, one hen was still exceptionally fierce, though her mate seemed to have lost nearly all his aggressiveness. In another case, a cock still chased intruders in mid-July.

It was formerly believed that the parent robins attack and drive away their fledged young from the territory; alternatively it was sometimes stated that the young attack and kill their parents. Neither statement is correct. Adult robins occasionally attack trespassing juvenile robins belonging to other pairs, but this is not often seen, and certainly they do not usually attack their own young. After becoming independent of the parents the young gradually disperse, but this is not due to parental hostility, and they wander unmolested about the territory. The parent robins also seem to disappear in late summer, but this is not because they have been killed by their young, but because they become very retiring in habits in preparation for the moult. Careful search shows that they are still there. The juveniles, of course, finally 'disappear' in late July and August, when they moult into the adult plumage. These facts are doubtless the basis for the alleged attacks of parent robins on their offspring or vice versa.

When the juveniles stake out territories on their own in early autumn it sometimes happens that a juvenile claims a territory adjoining that of one of its parents and so comes into conflict with it, and in the same way a former husband and wife may also fight, but previous family relationships have no significance in such encounters. Each robin is by then an independent individual, and one may suppose that it does not recognize the former members of its own family – at least, it does not base any action on such recognition.

A popular Christmas card shows four or even more robins perching happily together on a holly branch. No more inappropriate symbol could be devised for the season of peace and goodwill. Should the depicted incident occur in nature, furious conflicts would arise. From September onwards till May, the woodlands, parks, gardens, and hedgerows of England are parcelled out into a great series of small holdings, each owned by one individual or by a pair of robins. Any robin staking out a territory after September can do so only as a result of the disappearance of an owner or by forcible ejection.

When the owner of a territory disappears, the owners of the neighbouring territories expand into the vacant site almost at once, and often the ground is fully occupied within twenty-four hours. Such encroachments are not normally attempted by neighbouring owners while a bird is in residence, presumably because the owner's song continually advertises possession. Encroaching robins often begin cautiously with very quiet song from low down in the bushes. If no opposition is encountered, they gradually change over to loud song uttered from the tops of trees, and ownership of the new ground has now been claimed. Wandering robins without territories also start claiming territory in the same fashion, with quiet song from the bushes leading to loud song high up. If such a bird tries to claim territory in ground already occupied, it usually meets with a vigorous attack from the owner, and may abandon its attempt almost at once (cf. the instance described in Chapter 2). In other cases the newcomer makes a more sustained effort, and then a serious fight may result.

Possession of a territory seems vital to a robin. Without one, a robin cannot acquire a mate or breed. Hence it might be thought that while bluff-fighting could be effective in maintaining territories once these are staked out, fighting is bound to be more bloody when one robin tries to dispossess another. This, however, is often not the case.

In all, eight encounters were observed in which one robin made a sustained attempt to dispossess another of its territory. In six cases the procedure was as follows: The newcomer sang loudly from high in a tree. During the pauses in its song-phrases, the owner robin replied with equally fine song. The newcomer now took wing for another song-perch, whereupon the owner promptly took wing in pursuit. After a brief chase between the branches, in which the owner did not catch up with the newcomer, the latter settled and again sang loudly. The owner did not continue the chase (it could now easily have caught the newcomer), but settled also and replied with loud song. This formal procedure, of alternate loud singing and chasing, was continued all over the territory. In two cases, after ten and twenty minutes respectively, the dispute ended with the newcomer leaving the territory. More remarkable, in three other cases, after a rather longer period, the owner ceased to chase the newcomer, ceased to sing loudly, and, after remaining in the territory a little longer with quiet song, left it for good, and the newcomer took possession. Another such encounter lasted for two whole days, but unfortunately the birds were not previously ringed, so it is not known whether the original owner or the newcomer was eventually successful. Thus in six cases of fighting for the territory which is so essential to a robin, the affair was concluded not only without bloodshed but without any true fighting whatever, and, while the owner chased and occasionally struck the newcomer, the latter even when victorious did not attempt to strike the previous owner; it seemed just to wear it down by persistence; and 'tout finit par des chansons'. Such psychological fighting is one of the most remarkable facts of bird behaviour. The singing which accompanies these contests is the finest robin song of the whole year.

Occasionally, territorial ejections lead to more serious fighting. One morning early in April an unmated cock invaded the territory of the mated male next door. When I arrived both cocks were in fine song. At intervals the newcomer attacked the owner; this was not observed in other fights. The owner resisted vigorously, and the birds would then close with each other, grappling with their legs and pecking furiously with their beaks. Not infrequently they lost their balance as a result, and twice the birds fell to the ground at my feet, where they still continued to peck hard for a little before separating. The mated cock was now obviously getting the worst of matters and had lost a great many feathers from his chin and one side of his face. After two hours he ceased to resist, and fled from the attacks of the newcomer. He now kept mainly in the bushes and sang only quietly. The newcomer sang loudly from high up in the trees and began extending his flights in all directions until he came in contact with neighbouring robins, thus discovering the limits of his new territory. The hen took no part in this encounter, and after a while began to follow the newcomer about as if he were her mate. ('None but the brave,' etc.) The displaced cock remained in the territory for two more days, looking very battered. He kept to low cover, but occasionally sang mildly, and was chased by the newcomer when noticed. After this he disappeared and was never seen again. The hen successfully reared a family with the victorious cock.

Even in the above case the ejected robin was not seriously damaged. Similarly Burkitt, in three years' close study of robins in Ireland, never saw one kill or seriously injure another. Occasionally, however, one robin does kill another. The late Eliot Howard informed me of one case he witnessed, and two other correspondents sent me similar records, while in several other fights a robin has been described as seriously hurt. There are also a number of published instances, for example the following from *The Field* in 1884: 'On Sunday last I saw two robins fighting under my dining-room window in such a fierce manner that they astonished me. I watched them until one actually killed the other, and then, like a game-cock, continued to peck his victim. I then went out to examine the poor bird, and found both his eyes out and his skull quite bare, and the victor flew onto a branch close to me, and began to sing in the sweetest notes. I then left the dead bird where I found it, and before I could get into the house he was at him again, pulling him about and standing on him, and he actually pecked a hole in his side.' Several other cases of one robin killing another can be found in the bird literature, but, considering how frequently robins fight, cases of death are extremely rare.

Clearly, since fighting can result in one robin killing or injuring another, the more usual bluff-fighting is of great advantage to the species. It also has another advantage over real fighting. When fighting seriously, two robins sometimes grapple with each other and fall to the ground, and, so oblivious are they of their surroundings, they can sometimes be picked up, still grappled together. Hence they might easily fall victim to animals of prey. Indeed a cat has been observed to catch both of two fighting robins.

Since a breeding pair of robins do not normally tolerate any other robins in their territory, it may seem surprising that one pair should have bred successfully in each aviary without killing or even injuring the other pair which were present with them continuously throughout the breeding season. Frequent chases and attacks were observed, particularly in the earlier weeks, but these showed no signs of becoming really serious. These results, and some casual observations made in the London Zoological Gardens on African bishop-birds (which in the wild are extremely aggressive), suggest that enforced crowding of territorial birds, provided that the attacked birds have sufficient space in which to escape, causes a great decrease in their aggressive behaviour towards each other. The captive robins and bishop-birds* attacked each other much less fiercely and showed far less threat-posturing than they do in the wild. The main reason, as discussed in Chapters 12 and 13, would seem to be that aggressive behaviour diminishes in intensity against a particular individual as time goes on. The captive robins still sang and postured fiercely at wild

* This bird was called a 'bishop' until 1935, when the editor of the *Ibis* changed it to 'bishop-bird' on receiving my paper entitled *Territory and Polygamy in a Bishop*.

robins outside, and when a strange robin was introduced into the aviary they attacked it violently.

Apart from feeding and sleeping, the robin spends most of its life in singing and fighting, and so far as emotional intensity is concerned, singing and fighting seem to be much the most exciting events in its life. On the other hand, sexual behaviour occupies comparatively little time and is attended by comparatively little excitement. It has often been stated that the sex instinct is much the most powerful in animals, but, in the robin at least, sex is greatly overshadowed by fighting.

It might also be wondered how the robin can afford to spend so much of its life in fighting and warning off intruders. However, except during the time that they are feeding their young, and in winter when food is short, many birds seem to have a lot of spare time. The robin uses his for singing and fighting. The great crested grebe, on the other hand, was found by Julian Huxley to spend much of the summer in courtship, and the cirl bunting, watched by L. S. V. Venables, seems to spend most of its spare time doing nothing in particular, a dull bird to watch.

This chapter may be concluded with an account of an unnatural occasion for robin fighting. A student of robin behaviour could scarcely wish for a more pleasing awakening than that provided by robin song in the bedroom itself. I woke to see a robin perched on the top of the open window. The same occurred the following morning, so a watch was kept which revealed the cause of the incident. This robin had newly acquired a territory in the copse adjoining the bedroom, and, doubtless when searching for food, had discovered its reflection in the window glass. On sighting the apparent rival the robin flew to attack. The rival, of course, rose in the air too, the robin rose higher, the reflection followed, until the bird reached the top of the glass, when, of course, the reflection suddenly disappeared, and the robin celebrated its sudden victory in song from the sash.

This attack was continued for about two weeks against all the windows on the same side of the house, and gave uncritical people the idea that the bird was trying to get in. After a fortnight the attacks ceased. Similar incidents have been not infrequently recorded in natural history journals and in the press, particularly from robins, but also from a great variety of other species which fight in spring. Sometimes exceptionally fine displays are seen, because, unlike most wild intruders, the opponent reflection always fights back. It need not, of course, occasion surprise that a bird should mistake its reflected image for a rival. A bird may likewise mistake the reflection of its voice, Richard Jefferies recording a cuckoo which repeatedly cried back at an echo.

5

The Formation of Pairs

The birds that live i' the field
On the wild benefit of nature live
Happier than we; for they may choose their mates
And carol their sweet pleasures to the spring.

J. WEBSTER: *Duchess of Malfi* (c. 1614)

Tradition assigns St Valentine's Day for the pairing up of wild birds, which, since most British birds do not nest at least until the end of March, I used to suppose was much too early. But observations at Dartington showed that, so far from this being too early, the first robin pairs were formed in the middle of December, over three months before the birds nest, while by 14 February almost all the pairs had been formed. This association into pairs is quite distinct from courtship proper, which does not occur until just before nesting commences.

The time of pair-formation has been accurately deter-mined in few other resident British birds. Eliot Howard found that chaffinch, yellow-hammer, and reed-bunting form pairs in February and March, as probably do most other res-ident song-birds, but the blackbird apparently pairs up in late autumn, while some starlings form pairs in late autumn though others not till the spring. On the other hand, in most or all the migrant song-birds the males arrive at the breeding grounds shortly before the females and the pairs are not formed until April or May, often only a few days before nest-ing starts.

Darwin's theory of sexual selection states that the bright colours of male birds result from the choice exerted by the females in selecting their mates. 'The females are most excited by, or prefer pairing with, the more ornamental males, or those which are the best songsters, or play the best antics.' In the robin, as in almost all other birds so far studied, it is the female which selects the male. The unmated cocks sing in iso-lated territories and the hens come to them there. The cocks do not leave their territories in search of the hens. Often a hen does not take the first unmated male which she encounters. For instance, while a hen robin with an autumn territory often pairs up with the cock of an adjoining territory, she sometimes moves farther even when her immediate neigh-bours are still unmated. Also, a hen occasionally deserts her first choice for a second. Since in an average year about one-fifth of the cock robins fail to get mates, there is ample scope for the sort of selection which Darwin supposed to exist.

The song of the male robin, though important in fighting, also draws the attention of the unmated female to a potential mate, but no evidence was obtained during the present study as to whether the cocks with loud songs got mates more easily than those with poor songs, or whether the fiercer males, and those with the larger territories, got mates more readily than those which were milder or had smaller territories. Such evidence might be hard to assess. One point, however, did emerge, that a male deserted by its mate has often managed to acquire and retain a second mate, and that males unmated in one summer have been mated the next and vice versa. Hence chance, and perhaps also individual personal factors, play some part in pair-formation.

In her extensive study of the American song-sparrow, Mrs Nice found no evidence of female choice in regard to the plumage, song, belligerency, or size of territory of the males. 'Old females try to come back to their former homes; otherwise their "choice" of mates appears to be perfectly haphazard.' The last statement is probably too sweeping, since if it were correct it would be difficult to see how advertisement-song, territorial behaviour, and other presumed products of sexual selection ever came into existence. But these observations show that chance plays a large part in pair-formation, and that the presumed influence of female selection cannot be demonstrated in wild birds unless by a careful statistical analysis over a long period of years. It may be added that, in the robin and other territorial birds, pair-formation is the only stage at which sexual selection in the sense of female choice can occur, for owing to the existence of territories the female robin does not see the later courtship of any male except her own.

Robins form into pairs soon after it is first light in the morning. Hence persistent early rising was needed to witness this occurrence, and it was not till the fourth year that I was lucky enough to arrive just before proceedings began. Soon after dawn an unringed robin was noticed hopping about and occasionally feeding. Its actions suggested it was a hen. Gradually it made its way to the territory of an unmated cock some sixty yards away, feeding as it went, and eventually came to the edge of this territory. Uttering a brief song-phrase, the hen (as events proved her to be) now flew into the territory close to the unmated cock, who was singing on a bough. The cock stopped singing, flew up to the hen and postured aggressively, then retired and sang loudly. The hen uttered another song-phrase and then flew right up to the cock, and the cock again retreated and sang loudly.

From this instance and less complete observations on other pairs, the general procedure of pair-formation would seem to be as follows: The hen at intervals flies right up to the cock, who either retreats immediately or postures aggressively. In the latter case the hen sometimes postures back. The cock sings loudly and some hens also sing well, though other hens utter only a few phrases and some are not heard to sing. At intervals throughout the performance the birds break off and take, or appear to take, food from the ground. Such a humdrum break in otherwise excited proceedings has been noted by Eliot Howard in other phases of bird behaviour. After a period of this excited singing and sometimes posturing, the cock begins to follow closely after the hen, and the pair is formed.

The degree of excitement shown by the two participating robins has varied enormously. Some couples were hardly at all excited, while on the other hand one pair maintained their excited singing and posturing for two days instead of the usual one to four hours. So violent was the posturing in this latter case that it was first recorded as an unusual type of fight between two rival cocks. This was in the fourth year of observation, when it might be supposed that I was experienced, and shows how easily bird behaviour can be misinterpreted, especially in birds in which the two sexes look alike. Fortunately the newly arrived bird was trapped and colour-ringed at once. Otherwise I would never have realized that the supposed rival cock of the first two days was the same individual as the hen of the third day. This hen finally demonstrated her sex by laying eggs and raising a family.

Thus the first stage of pair-formation in the robin is characterized by both threat postures and the loud singing typical of aggressive behaviour. One critic wrote that this showed it was wrong to suppose that the robin's display was threat posturing, it was really courtship; or at least it must be admitted that such display featured in courtship as well as in fighting. As against this latter view, the postures during pair-formation are indistinguishable from typical threat postures. Further, courtship display normally continues up to, and rises to a climax at, coition. On the other hand, this posturing at pair-formation usually dies away after a few hours and is not seen between the pair again, while coition does not occur for another two or three months and is preceded by a different posture.

The most probable explanation is that, up to the moment of pair-formation, the unmated cock robin, and sometimes the hen too, has been living in an exclusive territory from which all other robins were driven out; now, however, there is a strange robin which must not be driven out. At first the cock, and often the hen as well, is not properly adjusted to this new situation, and when the mate comes too close shows traces of the now quite inappropriate aggressive behaviour. Gradually the pair quiet down as they become adjusted, and the singing and posturing die away. Support for this explanation is provided by the fact that those robins, both cock and hen, which showed most threat posturing and song during the first stage of pair-formation were also the fiercest in attacking trespassers.

Since all robins except the future mate are driven from the territory, the unmated cock robin must be able to distinguish a possible future mate from a trespassing robin. In plumage and size the cock and hen robin look alike to an ornithologist, and presumably to a robin too, but with practice at least some hen robins are distinguishable from cocks by slight differences in attitudes and behaviour, they look rather slimmer and act more nervously, and probably such differences are more apparent to a cock robin than to a human observer.

But the problem of the unmated cock robin is not that of distinguishing the two sexes apart, but the much harder one of distinguishing a hen robin in search of a mate from trespassing robins of either sex. In my original account I came to no conclusion on this point, but I believe now that the main factor involved is that the hen in search of a mate persistently flies right up to the cock and does not retreat when he postures, whereas intruding robins normally avoid the owning cock and leave the territory when attacked. On this view one might expect that, as observation shows to be the case, recognition of a potential mate is difficult and by no means instantaneous. This view cannot be considered proven, but it fits the facts for the robin. Similar behaviour occurs in the pair-formation of the common heron and American red-winged blackbird. In these species also, the unmated male starts by chasing or attacking an arriving female, and the female in search of a mate is distinguished by her persistent return to the territory, whereas other females move right off. In the heron, also, females may show aggressive behaviour, like female robins.

The problem of how the unmated male distinguishes a potential mate has been investigated in few other birds. Gentoo penguins apparently experience more difficulty than does the robin, since at the beginning of the season both sexes are said to behave similarly, and the pairs are formed only gradually.

On the other hand, there are species in which the problem seems to be simpler. For instance, the cock song-sparrow treats all strange hens alike, and the behaviour of the latter determines whether or not a pair will be formed. In this bird, unlike the robin, the cock simply has to distinguish the two sexes apart. They are alike in plumage, and sex recognition apparently depends on a special call given by the hen. In other species there are marked plumage differences between the two sexes, and experiments with stuffed specimens have shown that in several species the male immediately distinguishes a bird's sex by its plumage. Indeed when artificial black moustachial stripes, which are the male's most characteristic feature, were experimentally attached to a mated female American flicker, her mate promptly treated her as he would a rival male, and drove her away. There are other species, such as the red-necked phalarope, in which one sex appears not to recognize the other at a distance but does so immediately at close quarters.

While the male bird sometimes has difficulty in distinguishing a potential mate, the female's problem is usually much simpler, since in territorial species the unmated male claims an isolated area in which it sings loudly or displays conspicuously. How the female distinguishes a potential mate in the case of species which form pairs when in flocks has not been studied. Indeed, far more work is needed on the whole subject of pair-formation in birds.

To return to the robin, the first stage of pair-formation

with loud song and posturing lasts from an hour or two up to, in one case, two days. After this there follows a much quieter second phase, in which the hen takes short flights about the territory followed by the cock, who now sings very quietly indeed. Similar behaviour is recorded in the red-winged blackbird at the same stage. In the robin this phase lasts one to three days, and, as already noted, the hen frequently trespasses into neighbouring territories and gets chased out again, until she has learnt the boundaries of her mate's territory.

After this phase there follows a long 'engagement period', which often lasts several months. During this time cock and hen almost ignore each other. They are seldom seen together, the hen is rather retiring, and the cock returns to moderately loud song. The cock clearly recognizes his mate individually, since he does not chase her out as he would a trespassing robin, and observation shows that he can distinguish her at least thirty yards away, which indicates remarkably accurate powers of vision. Many other birds have now been proved to recognize their own mates individually. Jackdaws and swans recognize each other by facial differences, and in jackdaws and night herons, call-notes as well as appearance have been shown to be important.

A spell of extremely cold weather sometimes occurs after the time when a robin pair has formed. There are two records under these circumstances of the cock and hen robin separating and reverting to individual territories until the end of the cold weather, when they rejoined each other. This was not observed at Dartington, but the winters are mild there. An unusually cold spell is well known to inhibit breeding behaviour in many birds.

In robin pairs constancy for the breeding season is the rule. However, in four instances at Dartington a hen deserted her cock for another, three of these desertions occurring during the engagement period, before nesting, and the fourth between the first and second broods. In two cases the hen transferred to the cock of the next-door territory, and in the two other cases to a cock some three hundred yards away. Burkitt also found two hen robins deserting their mates between broods, and there is one record of both cock and hen of a robin pair acquiring different mates for their second broods – as Ben Battle said, 'the love that loves a scarlet coat should be more uniform'.

In a fifth case observed at Dartington a cock robin temporarily changed his hen, but later got her back. A hen arrived and paired up with a cock. The next morning a second hen arrived and repeatedly flew up to the cock, and for the rest of the morning these two exhibited behaviour typical of the first stage of pair-formation. Neither paid attention to the first hen, who did not come near them but wandered about occasionally singing. The following day the new pair showed behaviour typical of the second stage of pair-formation, the cock following the hen about and warbling softly, while the first hen behaved as before. No further visits could be paid for a week, and then the second hen was found to have disappeared and the cock had re-mated with his first hen.

In four other cases a hen disappeared soon after pair-formation, perhaps deserting. In all, some 10 to 18 per cent of hens deserted, and Mrs Nice found a similar proportion among song-sparrows, these, as in the robin, occurring either in the engagement period or between broods. In the American house-wren and blue-bird, in which there is much less attachment to territory, desertion between broods is much more frequent.

While such divorces occasionally take place, the robin

normally has only one mate at a time. Indeed monogamy is far commoner than any other state among song-birds, presumably because two parents are needed to raise the young. However, almost every rule of robin behaviour is broken by one or two individuals, and one cock robin was bigamous. The two hens held separate territories, from which each regularly drove the other if she attempted to trespass, while the cock ranged freely over both territories. In the middle of March the cock spent most of his time with one hen, who now built a nest. So soon as the first hen was incubating the cock went mostly with the other, and after a while she also was rarely seen about and was presumably incubating. When the young of the first nest hatched, the cock spent most of his time with them. Unfortunately, before the second hen hatched out her young, thus facing the cock with the formidable task of feeding two broods at once, he disappeared. It may be added that, within twenty-four hours of his disappearance, two new cock robins had arrived and were claiming with loud song the two halves of the territory; a striking illustration of how quickly unoccupied robin territory is filled.

Three correspondents have reported cases of bigamy in the robin. In one of these, it was observed that each hen had a separate territory from which she chased the other, and the cock was seen feeding both hens. Both hens nested, but one was taken by a hawk before her eggs hatched. In one of the aviaries the owning cock robin also nested with both hens, but here conditions were unnatural.

Occasional cases of bigamy are recorded in other normally monogamous European song-birds, such as the chaffinch and some of the warblers, while four such cases were observed in the American song-sparrow. In the American white-crowned sparrow bigamy is relatively common, each female defending her portion of the male's territory against the other female, using song and threat display like a fighting male. Two European song-birds, the corn-bunting and the wren, have been claimed to be polygamous, the former having up to seven hens, and polygamy also occurs in certain American and African song-birds.

In those cases in which the hen robin stays behind in autumn after breeding, she not infrequently takes up a territory adjoining that of her former mate. But the pair are now antagonistic and drive each other out if trespass occurs. In four cases at Dartington and two others observed by Burkitt, such hens paired up again with the same mates in the following spring. Indeed one Dartington pair mated together three years in succession, and it is interesting that although the hen was unusually aggressive on the first occasion, she was very mild on the next, suggesting that the birds had remembered each other. In one other case the hen moved into the same territory in three successive springs, the same cock being involved in the second and third springs.

As regards the migrant hen robins, one at Dartington, also one recorded by Burkitt, returned in a second spring and paired with the same cock as in the previous year. In another case a migrant female came to the same territory in three successive springs, but on each occasion found a different cock in possession. Similarly Burkitt observed two cases of a migrant hen returning to her former territory and pairing with the new male in possession. Hence the occasional cases of re-mating among robins may be due to a return to the former territory rather than to the former mate, and Mrs Nice came to the same conclusion for the cases of re-mating among song-sparrows. Not all hens return to their old territories. Three other hen robins which returned to Dartington in a second spring selected different territories, although in each case their former territory was held by an unmated cock. Indeed in one case the old territory was held by the former mate, but, although the latter was still unmated, the hen paired with the cock of an adjoining territory.

These results dispose of the view held by some nineteenth-century ornithologists that the robin pairs for life. It must, of course, be remembered that robins, like most birds, revert to a nearly sexless state in autumn and winter, so that each breeding season represents a fresh start. The proportion of robins which re-mate is probably as high as, if not higher than, that in most British song-birds, but detailed evidence is lacking in most species. There is also a record by Burkitt of a pair of robins which remained together during the autumn, instead of holding separate territories. This pair, but no others, might be described as pairing for life.

Among European song-birds the marsh tit has been proved to pair for life, and the same has been claimed for other tits, for the nuthatch, the stonechat, and some other species, but in most cases the evidence is quite inadequate. Among American species, each pair of wren-tits remain together in the same territory for life, and there are other species in which the pair remain at least loosely associated outside the breeding season. In the mocking-bird the pair usually separate and hold adjacent territories in autumn, but, as in the robin, there is one case of a pair remaining together in autumn. Outside the song-birds, life-pairing has been found or claimed in the crow family, parrots, moorhen, little grebe, geese, swans, and others. Hence while it is unusual, life-pairing probably occurs in a number of species, though instances among wild birds under natural conditions have only rarely been proved with certainty.

6

Courtship

Win her with gifts, if she respect not words.

W. SHAKESPEARE: *Two Gentlemen of Verona*

From a human standpoint the sequence of the robin's family life is curious. First the hens go out to choose their mates, the cocks awaiting them in their domains. After what seems like a brief fight the pair become engaged, and for the next two or three months they share a territory, but otherwise almost ignore each other, and there is no courtship whatever. Then the hen, unassisted by her mate, constructs the family home. About the time that this is completed mating occurs, very infrequently and with little excitement. Starting about the same time, but continuing for much longer and with much more attendant excitement, the cock courts the hen by bringing her food.

The term courtship has been used vaguely for three different phases of bird behaviour, first for the advertisement display of an unmated cock bird for a hen, secondly for the display which leads to coition, and thirdly for what may be termed bond-forming display.

Only in a broad sense can the advertisement song of the cock robin be described as courtship, since its object is not to stimulate the hen sexually, but simply to draw her attention to his presence so that a pair can be formed. This often occurs several months before the true courtship begins.

In the robin the second phase, sexually stimulating display leading to coition, is reduced to a minimum. Coition occurs only during nest-building and the laying of the eggs, and then only a few times each day. The hen stops still and lowers her head, somewhat humping the body, and the cock mounts. Normally there are no other preliminaries by either sex.

Very occasionally a cock tried to mount without the hen giving the preliminary signal, as once when a hen was inviting her mate to feed her, and on another occasion when the pair were feeding young. Occasionally, too, the reverse was observed, a hen giving the signal for coition and the cock failing to respond. Sometimes the hen then assumed an exaggerated humped attitude and lowered the head several times in jerks, also swaying the body from side to side; and after one such demonstration the cock mounted. Howard pointed out that the value of sexual display was to synchronize the actions of the two sexes, and that it was primarily when one sex failed to act appropriately that elaborate displays were seen. How the elaboration of a display enables a bird to do something of which it previously seemed incapable is, as discussed in Chapter 3, unknown.

When one of the aviary pairs was in coition the other

cock, which was not breeding, flew up, drove the cock off the hen's back and mounted in turn. The hen seemed unaware of any change. A similar interruption occurred on the only other occasion when coition was observed in the aviary. Parallel behaviour has been observed in the blackcock, the rook, and a number of other birds in the wild, but would be almost impossible in wild robins owing to the existence of territories.

An American worker placed a male dove in one cage and a female in another; the male displayed towards the female and later the female laid an egg, which was, of course, infertile. By this and other experiments it was shown that one function of courtship in doves is to stimulate the female to produce an egg; indeed it has recently been demonstrated that a female dove will lay if provided with a mirror, since she courts the bird in the mirror, which, of course, displays back, thus stimulating her to lay. It is not known whether courtship has a similar function in other birds, but it is related that William Harvey, discoverer of the circulation of the blood, had a tame parrot, assumed to be a male, which he was in the habit of stroking, as a result of which it eventually demonstrated its true sex by producing an egg.

In many song-birds the courtship period is characterized by violent chases of female by male, and at a later stage there may be elaborate display on the ground, in which both sexes may take an active part. But apart from a feeble swaying of the body noticed on two occasions only, the cock robin seems to have no pre-coitional display at all, while that of the hen is rare if her adoption of a motionless attitude is excluded. In this respect the robin seems almost unique among birds.

In some species of birds courtship ceases completely when the eggs have been laid, and it is not resumed unless the birds start another nest. But in other species courtship continues long after the need for coition is over. This post-nuptial or bond-forming display was first described by Julian Huxley in the great crested grebe, and later observation has shown that it occurs in many other birds. In the great crested grebe and others the display takes the same form as the earlier courtship, and indeed the male may actually mount the female, though no further eggs follow. In the robin, however, the bond-forming display is of an entirely different kind: the cock feeds the hen.

In courtship-feeding the hen robin utters a sharp monosyllabic call, then, as the cock approaches with food, she partly lowers her wings and quivers with excitement, while the call changes to a rapidly repeated note as the cock finally feeds her. The hen's attitude and calls are indistinguishable from those of a young robin being fed by the parents. The cock often feeds the hen over and over again, the latter begging insistently. Indeed, one of the hens in the aviary repeatedly begged the cock to feed her though she was perched on the food tray surrounded by mealworms while he was in a distant corner where there was no food.

Courtship-feeding does not lead up to other forms of courtship, and in particular it has no connection with coition. The birds derive considerable excitement from it, and it seems an emotional end in itself. Huxley suggested that the post-nuptial courtship of the great crested grebe was of value as an emotional bond between the pair, which is important in birds in which two parents are needed for rearing the brood. 'Their mutual passion, and the acquired knowledge that their joint labour is necessary to produce sustenance for their numerous family, induces the wild birds to enter into a contract of marriage,' wrote Erasmus Darwin in his *Zoonomia* in 1801;

though put rather anthropomorphically, this contains the essential idea. I formerly supposed that this sufficed also to explain the habit of courtship-feeding, in the robin and other birds. But later observations by other workers on the pied flycatcher and great tit have shown that the hen bird receives a substantial proportion of her food from the cock, both before laying and during incubation. Further, food is sparse at the start of the breeding season and as a newly laid robin's egg weighs 2½ grams, the hen robin has to produce 12½ grams in eggs, just over two-thirds of her own weight, within a very few days at a time when food is sparse. Hence any additional nourishment is probably of real value to her. Likewise if the cock feeds the incubating hen when she comes off the eggs for refreshment, she will be able to return quicker, and this is probably advantageous, as the brown back of the sitting hen matches the background much better than do the pale eggs. I therefore conclude that the food given to the hen by the cock is of real value to her.

Courtship-feeding starts in late March and continues during nest-building, laying, and incubation, also between broods, but diminishes in frequency after the young hatch, and ceases about the time that the hen stops brooding the young, when they are a week old. To the last, there was one exception. Just after her family of five young left the nest a hen robin lost her mate. She continued to feed the brood, and, three days later, acquired a new mate. The hen now played two rôles at once, the first feeding her young, the second begging from and being fed by her new mate in courtship-feeding. (The cock fed the hen but not the young, as described in the next chapter.)

Some other exceptional incidents may also be noted. On two occasions my presence near to a robin's nest containing young alarmed the parent robins, so that they would not come to feed their brood. The hen robin thereupon turned and begged for food from her mate, although her own mouth was full of insects for the young. Not infrequently when alarmed near the nest a bird performs some quite inappropriate action, but seldom is the inappropriateness so obvious as in this case.

A similar incident occurred near another nest. In this case the cock was holding a large caterpillar destined for the young, and when the hen begged from him he, seemingly automatically, put it in her mouth. Almost immediately afterwards he apparently recollected that this was not the caterpillar's intended destination, and he drew it back from the hen before she appeared to realize what he was about. She now begged him for food again, but this time the cock failed to respond, whereupon the hen tried to drag the caterpillar from him and a tug-of-war ensued. Unfortunately this was interrupted by a passer-by and the birds flew off.

G. J. Renier observed a robin pair which appeared to become estranged, and the hen would no longer accept the cock's proffered food. After a few days the hen did finally accept a worm, and the cock then returned with worm after worm in rapid succession. Such excessive activity after abnormal prevention is characteristic of instinctive emotional behaviour in birds.

In a wild state the cock robin fed the hen and the reverse was never witnessed, but in the aviaries a cock occasionally begged food from a hen. Further, after one of the hens had nested unsuccessfully, she began to carry mealworms about instead of swallowing them straight away, and on a number of occasions she offered them to other adult robins in the aviary. On one occasion a cock almost accepted a mealworm from her, and the other hen actually did so. While these actions are abnormal, they show that each sex possesses the potentialities of the behaviour of the other. This seems fairly widespread in birds. Aristotle was the first to record that, in the absence of a male, two female doves would form a pair, one of the females acting as a male, and this has now been observed in many other species. It is also known that, if isolated, the females of some birds will adopt male behaviour. Female birds can also be induced to behave like males through injection of male sex hormone, while in some species in which the male does not normally incubate the male can be made broody if injected with the hormone prolactin. Finally, there are a few species, such as phalaropes and button-quails, in which the sex rôles are normally reversed, display and bright coloration being typical of the female and the male being dull-coloured and carrying out nest-building, incubation, and the raising of the young. Hence the factors controlling 'maleness' and 'femaleness' seem different in different species of birds, and the whole subject would make an interesting experimental study in the interrelation of genetic, hormonal, and external environmental factors. It is not altogether irrelevant to note that nearly all those characteristics of behaviour which Western Europeans consider to be typical of one or the other sex in human beings have been found by anthropologists to be typical of the opposite sex in some other race of mankind.

'There are some who say that the raven and the ibis unite at the mouth... They are deceived by a false reasoning, because the copulation of ravens is seldom seen, but they are often seen uniting with one another with their beaks, as do all the birds of the raven family; this is plain with domesticated jackdaws. Birds of the pigeon kind do the same, but, because they also plainly copulate, therefore they have not had the same legend told of them.' Thus Aristotle in his *De Generatione Animalium*, which provides the earliest reference to courtship-feeding, a habit which is found in a large number of birds, including, as well as those already noted, many finches, nuthatch and tree-creeper, tits, parrots, gulls, terns, many birds of prey, and others. In many other groups of birds it is absent, while in a few groups it is not found in most species but occurs in one or two. For instance, it is absent in nearly all game birds and ducks, but occurs in the American bob-white quail and the Carolina tree-duck.

In many species, like the robin, the hen adopts the attitude and calls of the young bird being fed by its parents, and the cock feeds the hen in the same way that it feeds its young. Thus the habit of regurgitation, found in finches and gulls among others, is primarily an adaptation for feeding the young, but the male also regurgitates in courtship-feeding. In other species courtship-feeding takes a different form from the feeding of the young, being most spectacular in the harriers, in which the male flies above the female and drops the food, the female catching it in the air.

In many species, like the robin, the cock feeds the hen not only in courtship but also during incubation, and as already mentioned, the extra food means that the hen can spend longer on the eggs. There are other species of birds, such as certain cuckoos, in which the male does not feed the incubating female and courtship-feeding seems primarily associated with coition.

It is interesting that in all known cases except one it is the male bird which normally feeds the female. The exception is the button-quail, but in this bird most of the other sexual behaviour is also reversed, hence this exception serves only to emphasize the fact that the giving of food is primarily a male characteristic, the receiving of it a female one. This remarkable bird should be studied further.

In some birds, such as doves, courtship-feeding is often reduced to billing, and no actual food passes. As noted in Chapter 3, there is a general tendency for the display actions of birds to become formalized, and this provides another example. Presumably in doves the ceremony, not the food, is the important part of courtship-feeding. Billing was not seen in the robin.

It has been recorded that when two cock robins were captured and placed in a cage they repeatedly fought each other, but later one of them broke its leg, after which the other fed it regularly. This and a similar case in linnets are probably to be interpreted as a reversion to juvenile behaviour on the part of the injured bird, rather than as the adoption of female behaviour by a male. As a more definite example, a correspondent informs me that when he rescued and fed an adult oiled guillemot, after a few days it begged him for food. Courtship-feeding is not known in guillemots, so this was clearly a reversion to juvenile behaviour. More remarkable instances are those of an adult frigate bird and an adult brown booby, each possessing only one wing and each at least several years old, which were found alive at breeding colonies of these seabirds. Similarly an old blind white pelican has been found alive at a breeding colony. In none of these cases could the bird have obtained food for itself, hence it must have been fed for years by other members of the colony. Courtship-feeding is not known in these species, so presumably the birds had retained their juvenile behaviour. This is corroborated in some experiments on a captive juvenile American shrike. When the observer continued to feed it, its food-begging behaviour was prolonged for at least six months.

Hence it seems that juvenile behaviour may remain latent in adult birds, reappearing in some cases when they are injured and in other cases in courtship. The same may happen in mankind. Childish behaviour is not unknown in invalids; and Lysander once stole the impression of Hermia's fantasy with sweetmeats, a principle exploited in more commercial times by advertisements for chocolates. Apart from mankind, the best-known occurrence of courtship-feeding outside birds is in a group of dipterous flies, the *Empidæ*, in which the male presents the female with food as a regular part of the courtship. The male often wraps the presented food in a web, and in one species the food is dispensed with and an empty web is presented, which demonstrates, like the billing of doves, that what matters is the ceremony, not the food. It is remarkable to find behaviour so similar to that of birds in this group of insects.

7

Nest, Eggs, and Young

The Robin and the Wren
Are God Almighty's Cock and Hen,
Him that harries their nest
Never shall his soul have rest.

WILLIAM BLAKE: *Popular Rhymes*

In South Devon robins usually start nest-building in the last week of March. If the first nest is destroyed, another follows, while if the April brood is successful, there is often a second brood in May. Egg-laying is common from the end of March throughout April and May, but eggs in June are much less common, and nests with eggs in July are only occasional, though they occur. The breeding season is similar throughout England, but tends to start and finish a little later in the northern counties.

Nesting starts earlier in a warm early spring than in a cold late one, while if the weather is very mild the robin has been known to nest even in winter, though this is exceptional. There are twelve published records of nests with eggs in January, many more in February, also three in December, three in November, and one in October. Such out-of-season nests are rarely successful; the robins usually desert if cold weather follows the mild spell. However, a Norfolk pair successfully raised a brood hatched on 8 December, and an Irish nest contained fully-fledged young on 7 February.

A robin's nest is almost too well known to need description. Usually it is placed in a hollow in a bank or on the ground, and it is built largely of moss on a foundation of dead leaves, the cup being lined with hair. The bird is so common that one can scarcely help stumbling on at least one nest during the breeding season, but it is an altogether different matter to find the nest of a particular pair, and I fully agree with Burkitt's unexpected conclusion that it is one of the hardest of all British birds' nests to find. The nest is well concealed, there are a large number of possible places for it, and the birds are shy about visiting it when a man is near. Each year the attempt was made to find every nest of the colour-ringed robins, and each year some escaped, and one knew there had been a nest only when the parents were seen feeding their fledglings.

With luck the hen was seen carrying building material, but the building hen does not draw attention to herself and builds for only a few hours on each of about four days. During the laying period it is extremely difficult to find the nest, as the hen visits it only for the short periods necessary to lay; she lays one egg each day, usually between six and eight o'clock in the morning (sun-time). During incubation the hen sits closely, so that it is hard to find nests by flushing the sitting bird. However, since writing the first edition of this book,

I have realized that the incubating robin leaves the eggs once, twice, or even three times in the hour, and she is then fed by the cock. Having seen a cock feed his hen, it is not usually difficult to watch the latter back to her eggs – except that courtship-feeding also occurs in the period before the birds have a nest, in which case one's watch is vain. A robin's nest is most easily found after the young are a week old, as then the parents pay repeated visits with food. Most of the Dartington nests were found at this stage. Even then, though some were found after a few minutes, others took several hours, and a few were never found.

In the robin, like most other British song-birds, only the hen builds. Once when a cock came near the partly built nest the hen chased him away, as also reported by another observer, but two different observers have reported him bringing nest material. It is difficult to see why the cock should almost never help in building.

At intervals robins acquire notoriety by nesting in a jam-jar, a letter-box, an old boot, a pulpit, a human skull, or even a dead cat. A Birmingham pair started to build their nest in an unmade bed during the breakfast hour, and when the owner realized what was happening, the bed was left unmade, and the birds eventually reared a family from it. The record for speed goes to a Basingstoke pair. A gardener hung up his coat in the tool-shed at 9.15 a.m., and when he took it down to go off to lunch at 1 p.m. there was an almost complete robin's nest in one of the pockets. A pair at Walton Heath built in a wagon, which, just after the young hatched, had to be moved to Worthing. One of the parents travelled there and back with it, about a hundred miles each way, feeding the young en route. A similar occurrence has been reported from Northumberland. Presumably considerable territorial complications resulted on these journeys. More up-to-date records include nesting in a car, and even in an aircraft. It scarcely needs emphasizing that the robins which select such nesting sites are doing nothing odd. A robin places its nest in a crevice or hole, and those provided by human beings are as good as, but no better than, natural holes.

The houses of human beings are not the only ones that the robin may use. One Dartington nest was built into the old nest of a song-thrush, and correspondents have reported robins building in old nests of yellow-hammer and wood warbler, and in holes in trees excavated by green and great spotted woodpeckers. One robin actually ousted a redstart, another a nightingale, and a third a wood warbler, and took over the nest, although in the two latter cases the original owner had already laid two eggs in it. In another case a robin and a willow warbler were found, each sitting on six eggs in a nest built by the latter species, and another pair raised a combined brood with a pair of pied wagtails. There are two other cases of a robin and a redstart laying in the same hole, and four of a robin and a great tit laying together. In at least two of the latter cases, the robin deserted. In another case a blue tit ousted a nesting robin, laid her own eggs among those of the robin and eventually raised a mixed brood of both species. Another robin was turned out of an old kettle by a wren. Yet another pair built in an old swallow's nest and had hatched four young when the swallows returned from the south. The swallows drove off the adult robins, and each of their young was found dead with a wound over the eye like a peck. Another pair of tree-nesting robins had their clutch destroyed by starlings. Although so pugnacious towards members of its own kind, in nesting disputes with other species the robin often comes off worst. Hence it is not altogether surprising that most British robins nest on the ground. The danger from other hole-nesting birds is evidently greater than that from ground vermin. When a British robin does nest off the ground, on a wall or a tree, it usually chooses a covered niche rather than a true hole of the type popular among tits.

Normally a bird has no difficulty in recognizing where it has started to build its nest, because the selected hole has individual characteristics. But this was not the case with a pair of robins found building in a group of pipes stacked on their sides. The birds evidently became confused as to which hole they were using, and placed nests or parts of nests in twenty-three of the pipes. The same site was occupied in two later years, when rather fewer nests were built. Another pair built a series of nests in regularly spaced gaps in the brickwork round a sewage filter, laying four eggs in one nest, two eggs in the next, and one in a third, while other nests were left uncompleted. A third pair of robins started to build in some pigeon-holes in a workshop. They built nests or parts of nests in twelve out of the sixteen 'pigeon!' holes, but eventually mastered the situation and successfully raised a brood in one of them. Such cases of multiple-nesting are not uncommon in other hole-nesting birds and need not occasion surprise, since bird behaviour is adapted to natural conditions, and a set of symmetrical holes is not a normal circumstance of their lives.

The British robin normally builds a cup-shaped nest, and uses dead leaves as a foundation, but not elsewhere. In this connexion it is interesting to quote William Turner, the first British ornithologist, who wrote in 1544. The following is Ray's translation in his *Ornithology of Francis Willughby* of 1678. 'Of the manner of building its Nest thus Turner from ocular inspection. It makes its Nest among the thickest thorns and shrubs in Spineys, where it finds many Oaken leaves, and when it is built covereth it with leaves, not leaving it open every way, but only one passage to it. On that side also where the entrance is, it builds a long porch of leaves before the aperture, the outmost end whereof when it goes forth to seek meat, it shuts or stops up with leaves. What I now write I observed when I was very young; howbeit I will not deny but it may build also after another manner. If any hath observed another manner of building let them declare it, and they will very much gratifie such as are studious of these things, and myself especially. What I have seen I have candidly imparted.'

That Turner, good observer that he was, did not altogether trust his boyhood observation is evident from his concluding remarks. Other writers were less cautious, and Turner's observation was quoted as a general nesting habit of the robin by Buffon and Bewick, as well as by many less distinguished ornithological writers of the eighteenth and early nineteenth centuries. The latest reference appears to be that in the last edition of R. Mudie's *The Feathered Tribes of the British Islands*, in 1878, though the story was refuted by J. Rennie in his *Architecture of Birds* in 1831. Apart from the persistence of some of Aristotle's and Pliny's errors into the eighteenth century, this is probably an ornithological record, and is the more remarkable as Turner carefully stated his doubts and that he had observed such a nest only once.

It is still more remarkable to find records in recent bird literature which show that Turner's observation, while admittedly highly unusual, may after all have been true. The British robin normally builds an open cup-shaped nest, but, when it is placed in a hole or in grass on a bank, it is occasionally domed, and there is sometimes a tunnel to it in front. Two observers reported a robin paving such a tunnel with dead leaves, and another nest in 'a hollow in a tree was partly screened with dead leaves. Again, there are several records of robins nesting in thick hedgerows or in conifers. Most such nests are open, but one was 'nicely domed'. Finally there is a record from Cumberland in which a robin covered its eggs with oak leaves during the laying period, removing the leaves when the fifth egg was laid and incubation started. Between them, these records cover the essentials of what Turner observed. It is unusual for a robin to conceal its eggs during the laying period, but another case is on record, from Cornwall. In this case the eggs were hidden in the nest lining until incubation started.

In rural England the robin usually lays a clutch of five eggs, sometimes only four or six, and occasionally as many as seven. Rarely, clutches of eight to twelve eggs are found. The latter sometimes contain eggs of two distinctive types, showing that two hens have laid in the same nest, but some of the eight- and nine-egg sets have been laid by one hen. The record robin clutch occurred in 1944, when a robin laid twenty eggs. Indeed, she might have laid more, but deserted when disturbed by a cat. Towards the end, the eggs lay three deep in the nest. This is quite abnormal.

In many birds, removal of the eggs as laid induces the bird to lay more, until the normal clutch is complete. A flicker (American woodpecker) was in this way induced to lay seventy-one eggs in seventy-three days. In other birds, such as the herring-gull, removal of one or more eggs does not induce the laying of more, unless the whole clutch is taken, when the bird starts a second nest elsewhere. In this matter there is variation not only between species but between individuals of the same species. Thus by removing part of the clutch as laid, one writer was able to make a robin continue to lay up to ten or more eggs. On the other hand, a correspondent writes that when one and two eggs respectively were taken from incomplete clutches of four eggs, the robin in each case laid only one more egg and then started incubation, *i.e.* she laid the normal number of five and no more.

The equally interesting reverse experiment, of adding eggs to a nest to see if the robin will lay fewer than usual, has not, apparently, been attempted (because the above experiments have usually been due to egg-collectors). However, when two robins have laid in the same nest each hen has laid a full clutch of four to six eggs, making a total of nine to twelve eggs, hence it seems probable that the laying of the first five eggs is not influenced by the number already present. There is a Cornish observation to the same effect. A pair of robins deserted their nest after they had laid three eggs in it. When the cold spell was over, the hen returned and laid not two more but five more eggs in the same nest, which now, therefore, contained the unusual number of eight eggs. The only detailed experiments so far performed on any song-bird, to see whether adding eggs to the nest causes fewer to be laid, showed that in tricoloured redwings such additions often had no effect, but sometimes caused fewer eggs to be laid.

Since, if they are removed as laid, a robin is at times able to lay ten or more eggs one after the other, why does it usually stop at four, five, or six?

> Coo-oo, coo-oo,
> It's as much as a pigeon can do
> To maintain two;
> But the little wren can maintain ten,
> And bring them all up like gentlemen.

The Gloucestershire folk rhyme suggests that the usual number of eggs laid by each species is determined by the number of young which the parents are able to feed.

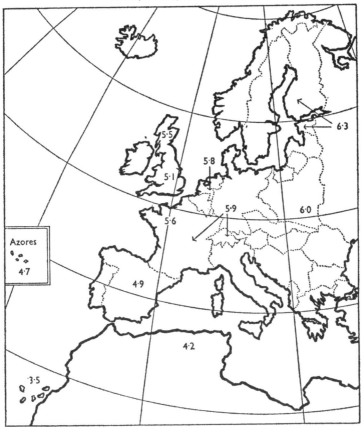

Fig. 3. Average clutch of robins

As shown in Fig. 3, the average number of eggs laid by the robin for its first brood varies from about 3.5 in the Canary Islands to about 6.3 in Scandinavia and Finland. It also varies with the time of year. In Germany and Norway, the robin usually lays six eggs for its first brood in May, but only four to five for its second brood in July. In England the seasonal change is less marked, but the proportion of clutches consisting of six eggs, as compared with five, increases steadily from March to early June, after which it declines.

Many other song-birds show similar variations in clutch-size, suggesting that the underlying causes apply widely. That clutch-size is larger in May and June than in April or July, and larger in the north than the south of Europe, suggests a connexion with the hours of daylight, as the day is longer in midsummer than in early spring or late summer, and longer in the north than the south of Europe. A longer day gives the parent robins longer in which to find food, hence it may enable them to raise more young at one time. But differences in day-length cannot account for all the variations. For instance, May and June clutches of the robin and other birds are larger in Central Germany than in England, but there is no appreciable difference in daylength between the two countries. The British insect fauna is poorer in species than that of the Continent, but it is not known whether the quantity of insects available for young birds is also less.

In the Cornish nest with eight eggs mentioned previously, the parent robins managed to raise eight young, but towards the end they are said to have looked very emaciated, hence in a less good season they might well have failed. Another correspondent reports a July nest of the robin in which the cock would not take his proper share in feeding the young. With only one parent on the job, four of the five young died in the first week. Such incidents suggest that the number of young for which two parents can find enough food is strictly limited. To conclude, there is a case for thinking that clutch-size is adapted to the average maximum number of young for which the parents can find enough food, though the evidence is not as yet by any means complete. Here is an interesting problem for future research.

The eggs of the robin are white, speckled with small reddish-brown spots. While this is the commonest type, other varieties occur. The background may be plain white, pale bluish, or pale yellowish. The spots may be small and faint, or small and dark, or there may be clear-cut blotches, or fine hair-like lines. The markings may be pale yellowish red, or dark brown, or grey, or purple. The spots may be so crowded that they hide the ground colour, or crowded at the blunt end and more spaced at the other, or forming a well-defined cap at the blunt end, or confined to a ring round the blunt end, or sparse all over. They may even be absent, in which case the egg may be either dull white or highly polished. Usually all the eggs in a clutch are similar to each other, but sometimes one is lighter than the rest. Egg-collectors have shown that each hen lays eggs of the same individual type throughout its life, and even when one egg differs in colour or size from the rest of the clutch, this peculiarity tends to be repeated in later layings of the same bird.

Eggs of an unusual type, with a faint bluish tinge and minute red spots, were found in a Kent orchard in the years 1909, 1912, and 1922. Since the average life of a robin is only one year (see p. 122), it is highly improbable that the three sets were laid by the same bird. It may therefore be suggested that the last set was laid by a descendant of the individual which laid the first. Likewise in Essex one year, three robins nesting near each other laid pure white eggs of an unusual type. Presumably these three hens were descendants of the same abnormal ancestor. Such cases suggest that egg-colour is inherited in birds, a fact which has been proved for brown as compared with white eggs in poultry, and which has commercial importance owing to the popularity of brown eggs among breakfast-loving Englishmen; but among

New York housewives this colour prejudice is reversed.

In many birds which nest in the open, the eggs are coloured to match their surroundings, thus concealing them from enemies. In many birds which nest in holes, the eggs are white, which makes them visible to the parents in dim light. Except for these cases (and parasitic birds, mentioned later), no value has been suggested for the colour of birds' eggs. Yet if their colour were useless, one might expect far less constant differences between the eggs of different species, and much greater variation among individuals of the same species, than actually exist. That eggs are potentially variable is shown by the abnormal robins' eggs already mentioned, and by the large number of unusual varieties in oological cabinets. But even before the attentions of egg-collectors became serious, such varieties were rare, which suggests that they must be eliminated by natural selection.

I became so intrigued with this problem that, some years after completing the third edition of this book, I made a survey of the colour of the eggs of all known species in the large thrush and chat family. This showed the existence of the two tendencies already mentioned as holding for birds in general, namely for white (or pale-blue) eggs in hole-nesters like the wheatear or redstart, and for darker and browner colours which help concealment in those species which build open cup-shaped nests. Of the latter, the chats which nest on the ground usually have dense brown spots which help the eggs to merge with the background, while the thrushes which nest in bushes usually have blotches or mottlings, which help to break up the outline of the eggs when viewed from above. White eggs faintly or moderately spotted with reddish brown, like those of the robin, proved to be typical of the species which resemble it in nesting in covered niches, this pattern evidently being evolved as a compromise between two con-flicting needs, a white background helping the parents to see the eggs in a dim light, and spots helping to conceal the eggs from passing enemies. Much harder to account for are the blue eggs of those members of the thrush family which nest in deep shade, but this is presumably connected in some way with concealment, in view of the nature of the light which penetrates to their nests.

One other British bird sometimes lays a white egg with reddish spots, namely the cuckoo, but this is for a different reason. It has now been abundantly proved that the eggs of the cuckoo tend to match those of the species which fosters its young, natural selection operating through the foster parent recognizing cuckoos' eggs which are too unlike its own, and either ejecting them, building over them, or deserting. In Britain, the cuckoo has not evolved a particularly robin-like egg. However, though individual cuckoos exist which confine their attentions to robins, the robin comes only fifth on the list of cuckoo fosterers, being duped less often than meadow pipit, reed warbler, pied wagtail, or dunnock. In parts of Central Europe the robin is the commonest host, and here the eggs of 'robin-cuckoos' show a decidedly closer resemblance to those of the robin than is the case in England. In Moravia, of ninety-two robins' nests found with a cuckoo's egg, 14 per cent were deserted, so selection by the robin is evidently stringent.

The hen robin usually starts to sit as soon as the clutch is complete, and incubation lasts thirteen to fourteen days. If the eggs fail to hatch, the hen may continue sitting for long past the usual time, in one recorded case for five weeks, and in another for forty-eight days, but eventually she deserts. The cock does not normally incubate.

In the robin, as in other song-birds, incubation is far from continuous during the daytime. At times the hen leaves the eggs of her own accord, and at others the cock approaches and calls her off with a short song-phrase. When off the eggs, the hen repeatedly calls the cock to feed her, which he does. If the cock is collecting more food when the hen returns to the eggs, he may bring a final mouthful there, but except under these circumstances, he does not often feed her when she is actually incubating, presumably so as not to reveal its whereabouts to possible enemies.

The hen robin sits closely, and is difficult to see as her brown back matches the surroundings. As pointed out by H. B. Cott, close sitting is characteristic of the birds in which the hen is protectively coloured but the eggs are conspicuous. When flushed, the hen robin usually flies straight away from the nest. One correspondent, however, observed a hen fluttering along the ground in the display formerly termed 'injury-feigning', but perhaps better known as 'distraction-display'. Other observers have occasionally seen mild exhibitions of the same type, but it is much rarer in the robin than in many ground-nesting birds.

When a man is near the nest, the parent robins usually hop round uttering a soft 'peep'. A high-pitched drawn-out 'see-ees' and a harsh rattling 'tic-tic-tic-tic' are also used in alarm, the 'see-ees', which is extremely hard to locate, particularly when a hawk is near, and the 'tic-tic' particularly when a rat, weasel, or other ground predator is near the nest, and the robins perch a few inches from it, calling violently and apparently distracting it. The above calls are not confined to alarm. For instance the 'tic-tic' call is sometimes used when one robin attacks another, and is also heard between robins preparatory to retiring for the night. To complete this list of the robin's call-notes, there are seven others, a soft 'sip' on migration at night, the invitation-to-be-fed call of hen and fledglings, their rapid vibrated call when fed, the squealing alarm of the fledgling when handled, a sharp 'tshrick' used when the adult launches an attack, a high-pitched variant of the song used in threat posturing, and the song itself. The song, like the 'tic-tic' call, features in several situations, including advertisement of territory against rivals and for unmated hens, threat-posturing, and the calling of the hen off the eggs. With so limited a vocabulary, it is perhaps not surprising that some robin calls should have several meanings. It is more curious that one situation, such as alarm or fighting, should evoke three or four calls, though their connotations seem somewhat different.

It is interesting that the rattling 'tick-tick-tick' used against a mammal and the ventriloquial 'see-eep' used against a hawk are extremely similar to the calls of a blackbird or song-thrush under the same circumstances, except that they are higher in pitch. But they are not like the alarm calls of a redstart, nightingale or bluethroat, which utter a guttural 'churr' or a melodious 'hweet'. Similarly the harsh 'tsreek' call of the fledgling robin sounds like a high pitched version of the corresponding call of a fledgling blackbird or song-thrush, and is unlike the call of a fledgling nightingale or redstart. Again the soft 'sip' given by migrating robins at night is like that of migrating song-thrushes. I therefore think it possible that the robin is not, as usually supposed, a 'chat' related to the nightingale, bluethroat and redstart, but a diminutive thrush which has taken to a chat-like way of feeding. Otherwise I do not see why it should call like a thrush, not a chat.

When the young robins hatch, the parents remove the eggshells, but any unhatched eggs are usually left in the nest.

The change from incubation to feeding behaviour occurs gradually. In the first few days the hen spends much time brooding the young, but this gradually diminishes, and she rarely broods by day after the young are a week old. When the cock comes to the nest with food for the young and finds the hen brooding, he often passes the food to her, and she passes it to the young.

When twelve to fourteen days old, the young robins completely fill the nest. If danger threatens, they flatten themselves and crouch down, and their dark plumage merges with the shadow of the nest cavity, while the pale flecks suggest small points of sunlight coming through the overhanging vegetation. This protectively coloured plumage becomes more important when the fledglings leave the nest.

The young were usually ringed when six to nine days old, as by then the leg is fairly well grown. It is dangerous to wait much longer because, though the young normally leave the nest after a fortnight, they may do so up to two or three days earlier if disturbed, and are then so helpless that they fall an even readier prey than usual to cats, rats, and other predators. Normally this scattering of the young is probably of value, since disturbance to a nest full of fledglings is most likely to be due to a predator, and if the fledglings scatter, losses are diminished. Several times when ringed rather late, one or more young robins have promptly left the nest when replaced in it. When this happens they can sometimes be induced to remain by replacing them in the nest and covering them with the hand till they quiet down, and after this treatment one restive brood remained a further three days in the nest. Sometimes one fledgling is particularly restless; this should be placed at the bottom of the nest, with the others on top of it (one of the fledglings has to occupy this position anyhow), and the extra difficulty in moving has then sometimes sufficed to keep it quiet, and to keep the brood in the nest another day.

Cock and hen robin recognize each other individually, and, to judge from the proportion which desert when a cuckoo's egg is placed in the nest, they are probably fairly sharp in detecting a strange egg. It therefore seems surprising that for feeding purposes the parent robins apparently fail to distinguish their own fledglings from those of other pairs. For instance, a hen robin was feeding her young which had left the nest two days. A rather larger fledgling with a longer tail, which had left the nest at least four days, approached from the neighbouring territory, and the hen of the first brood promptly fed it several times. A similar incident occurred in one of the aviaries. A few minutes after the incompletely feathered and tailless young had left their nest, a wild fledgling robin some fourteen days out of the nest, fully feathered and strong on the wing, perched on the aviary roof. The aviary parents flew up to it, at which the wild fledgling begged for food, and the aviary parents promptly tried to feed it through the wire, temporarily ignoring their own young.

It is probably the food-begging behaviour of the strange bird which elicits the adults' feeding behaviour, as shown more strikingly when a German ornithologist introduced an un-fledged linnet into a cage with some robins. When it gaped, the robins fed it. Again, a pair of wild robins which had lost their own young helped to feed a brood of nestling song-thrushes; there is a second instance of a robin feeding young song-thrushes, and a third pair repeatedly fed a newly fledged blackbird. Such offices may be mutual, for after raising two of its own young, a hen blackbird continued for two or three weeks to offer food to any bird which came near, and an adult robin was among those which accepted. Another pair of robins fed a fledgling cuckoo raised in the first place by meadow pipits, and in yet another case a pair adopted two fledgling wrens. The last 'is one of the best ornithological stories I have ever heard,' said Goethe. 'I drink success to you, and good luck in your investigations… If it were a fact that this feeding of strangers was a universal law of nature, it would unravel many enigmas, and it could be said with certainty that God pities the deserted young ravens that call upon Him.'

The food-begging behaviour of the young normally elicits the parent robins' feeding behaviour only when the latter are at the appropriate stage of the breeding cycle. This was well shown in the aviaries, in which the fledgling robins begged for food indiscriminately from all four adult robins, but only their parents, and not the other two adults, responded by feeding them. Again, in the wild, a juvenile trespassed into the territory of a cock which did not have young. The latter started to attack the fledgling, at which the fledgling, rather surprisingly from the human standpoint, begged it for food, but the cock did not feed it and continued to attack. In another case, referred to in the previous chapter, the male of a pair died just when the young left the nest. The hen continued to feed the young, and three days later she also acquired a new mate. The latter repeatedly fed the hen in courtship-feeding, but never responded to the food-begging of the young by feeding them. The parent robin is perhaps exceptional in responding to the food-begging of young birds only when it is at the appropriate stage of the breeding cycle. In quite a number of other species there are records under natural conditions of individuals which were not parents helping to feed fledged young, and in the swallow and moorhen the young of the first brood sometimes help to feed later broods.

Possibly some parent robins come eventually to recognize

their own fledglings, for G. J. Renier reports a cock robin collecting food for its fledglings near to the fledgling of another brood. The latter begged for food, and the parent robin 'absent-mindedly' started to put food down its throat, but then 'as if recollecting itself' it withdrew the worm and flew off with it to its own young. So perhaps a wise father knows his own child. Also, while a robin occasionally attacks a strange fledgling, I never saw a parent attack its own young. However, a correspondent reported one possible case of this by the male parent when the young had left the nest a fortnight; this was evidently an individual aberration, and the hen continued to feed the young. It seems that, with a few exceptions to both statements, the robin is normally able to recognize its young as regards refraining from attack, but is usually unable to recognize them for feeding purposes. Such inconsistencies in a bird's powers of recognition are discussed in a later chapter.

Since a fledgling robin begs from any adult robin which comes near, it acts as if it could not distinguish its parents individually from other robins. Indeed it does not even distinguish them as robins, since, as described in Chapter 12, begging behaviour can be elicited from a fledgling by presenting it with a bundle of red breast feathers.

Several ornithologists have experimented on the recognition of their eggs and young by black-headed and herring-gulls. These birds, unlike the robin, will accept as eggs almost any more or less rounded object of about the same size as an egg, including an india-rubber ball, a watch, a tin, or even a square brick with the corners roughly rounded, and these can be of the most varied colours. From this one might conclude that gulls had much less accurate vision, or were much stupider, than robins. On the other hand, gulls distinguish their own young individually from all others in the gullery after only a few days, from which one might conclude that they had much more accurate vision or were more intelligent than robins, which usually fail to distinguish their own young. Actually, the visual powers of both robin and gull are excellent, since both can distinguish their mates individually at considerable distances. The comparative failure of the gull to recognize its eggs and of the robin to recognize its young is related to the ways of life of the two species. There is considerable survival value to the robin but none to the gull in detecting egg-substitutes, since the robin is parasitized by the cuckoo, but in nature strange egg-like objects are unlikely to get into a gull's nest, and it does not usually matter if they do. On the other hand, while owing to the territorial system young robins rarely come into contact with the young of another pair, this frequently happens to young gulls in the crowded gullery, hence there is considerable survival value to the gull but not to the robin in distinguishing its own young. Bird behaviour tends to be adapted, as in this case, to the particular way of life of the species.

After they leave the nest, the young robins gradually acquire independence. This could be studied particularly well in the aviaries. A fledgling was first seen to pick up some food for itself eight days after leaving the nest (*i.e.* twenty days after hatching), and live mealworms were taken independently on the sixteenth day (twenty-eight days after hatching). On the nineteenth day after leaving the nest the fledglings were taking as much food on their own as they received from the parents. On the twentieth day the parents were taking mealworms, hopping about with them, and then, as often as not, swallowing them themselves instead of feeding them to the young, and the young were also begging for food much less than before. In both aviaries the parents ceased to feed the young on the twenty-first day after they left the nest, when the young had hatched thirty-one days.

To sum up a typical nesting cycle, building takes four days, laying five days, incubation fourteen days, fledging fourteen days, and the young become independent in another twenty-one days, a total of fifty-eight days or eight weeks from the day when the nest was started. No observations seem available for other British song-birds on the length of time taken by the young to become independent after they leave the nest. In the song-sparrow the corresponding period is eighteen to twenty days, and in the curve-billed thrasher, another North American bird, it is twenty-two to twenty-six days.

If the first nest is destroyed the parent robins soon start another, while if they raise their young quickly there is not infrequently time for a second brood. In this case the hen usually deserts the young of the first brood about the time that they are fledged, though sometimes up to at least three days before, and for the next three weeks the first brood are fed exclusively by the cock, while the hen builds and incubates for a second time. When the second brood of young has hatched, the cock is ready to help to feed them, as the first brood is now independent. When the second brood has fledged both parents normally stay with them; and this also applies to those first broods which are fledged too late to be followed by a second brood. There are also three cases on record in which a pair of robins successfully raised three consecutive broods in a season, and one of these pairs actually hatched out a fourth brood, though this last was unfortunately destroyed by a cat. In another case a hen robin, after one unsuccessful and then two successful broods, also left her second successful brood to the care of the male and started on a fourth nest. She raised this fourth family nearly to fledging, but they were then destroyed.

The female robin's habit of leaving the young of the first brood about the time that they are fledged, or rather earlier, is not uncommon in European birds which raise two broods. Obviously this habit saves considerable time. It is found primarily in species in which, like the robin, incubation is carried out exclusively or at least predominantly by the female.

This chapter has been concerned with what seems in many respects to be the dullest part of the robin's life. This is only through familiarity. The change in one fortnight from a speck in an egg to a blind, gaping, naked nestling, and in another fortnight to a feathered, tailless fledgling is as dramatic as any in the bird's life and presents many unexplored problems, problems which, however are more suitably investigated in the laboratory than in the field.

8

Migration

It is said to be a migrative species, but from no other
reason than their more frequent and numerous appearance about our habitations in winter.

G. MONTAGU: *Ornithological Dictionary* (1802),
under Redbreast

Migration is obvious in nightingale, common redstart, or
swallow, which reside in Britain only for the summer and in
Africa only for the winter. For robins in South Devon the
problem is less easy. Montagu, who lived in South Devon,
gave the answer quoted above, and at first I was of the same
opinion, particularly when census work showed that approximately the same number of robins reside at Dartington in autumn as in spring. Proper analysis led to a different, and more
complicated, conclusion.

Colour-ringing soon showed that all, or almost all, those

male robins which have once established territories are non-migratory for the rest of their lives. A few shift their territories, but such moves are extremely local and have nothing to do with migration.

In only three male robins at Dartington was there a possibility of migration after they had become resident. One cock appeared to be a summer visitor, arriving at Dartington in February, disappearing in autumn, and returning to its former territory the following February. Two others appeared to be winter visitors, arriving in autumn, leaving in early spring, and reappearing in their old territories in the following autumn. The breeding-ground of one of these apparent winter visitors was found, just 400 yards from its winter territory. Whether the two other apparent migrants moved only locally or were genuine migrants is not known.

Burkitt's study of ringed robins at Enniskillen in Ireland[*] shows that here also, once a cock had established territory, it normally remained, though as at Dartington there was a small amount of purely local shifting. Likewise one colour-ringed male robin at Oxford was apparently migratory, as it left a garden after breeding and reappeared there the following February.

At both Dartington and Enniskillen most male robins first appeared in autumn, and many of these birds were juveniles. Hence there are undoubtedly many males which never migrate during their lives. However, at both places a small number of males first appeared in January and February. Possibly these were juvenile birds which migrated after the summer and had now returned. That, as already noted, one of these

[*] Burkitt's account is repeatedly referred to in this chapter. Figures quoted later were not set out by him but are extracted from his observations.

males should have disappeared the following autumn and reappeared the next spring suggests that it may have migrated in two winters. But further data are needed before this possible migration of some first-year and possibly a few older males can be considered proven, since the males arriving in early spring might have moved only locally.

The situation is quite different in female robins. About a quarter of those at Dartington stayed throughout the year, with occasional shifts of territory. One of these definitely arrived as a juvenile in her first autumn, and stayed three years. On the other hand, the other females, 70 per cent in any one season, first appeared at Dartington between January and March to pair with resident males, and disappeared after the breeding season. A few reappeared in a following spring, but despite careful search only one was ever seen in the interval, this bird being found in late November a few hundred yards from its old territory. Burkitt's Irish robins were similar, nine females being non-migratory, one apparent migrant moving only 400 yards, and the other fifteen disappearing completely for the non-breeding season.

The above data prove that most female robins disappear after breeding until early in the following spring, but do not show whether this is a purely local scattering or a genuine migration. For long I inclined to the idea of a local scattering, being unduly influenced by the one apparently migrant female at Dartington whose movement was found to be purely local. The fact proving that some Dartington robins must migrate had confronted me for some time (as it likewise has the reader!) before its significance was appreciated. To repeat, census work showed that approximately the same number of robins resided at Dartington in autumn as in spring, and the same was true at Enniskillen. (See Table 1 overleaf.)

Table 1: Number of robins resident in part of Dartington area

	1935	1936	1937	1938	Average
In April	12	10	15	11	12
In November	13	13	12	10	12

Number of robins resident at Enniskillen

(The counts refer to rather different areas in the different years)

Spring 1923 14-16	Spring 1924 20	Spring 1925 26	Spring 1925 31
Dec. 1923 13-14	Dec. 1924 19	Dec. 1924 24	Dec. 1925 22

Table 2: Time of pair-formation among robins

| | AT DARTINGTON | | AT ENNISKILLEN | |
No. found forming pairs in:	Resident females	Migrant females	Resident females	Migrant females
Second half December	3	—	—	—
January	8	11	6	—
February	—	9	—	5
March	—	2	—	7
Early April	—	—	—	3

Table 3: Sex ratio among robins

AT DARTINGTON

	% of males	% of females	% unde-termined
On 1 April (based on 116 birds in four seasons 1935-8)	57	43	0
On 1 December (based on 77 birds in three seasons 1935-7)	60	21	19

AT ENNISKILLEN

	% of males	% of females	% unde-termined
Spring (based on 65 birds in three seasons 1923-5)	55	45	0
December (based on 57 birds in three seasons 1923-5)	74	18	9

The slight excess of males over females in summer seems fairly normal among song birds.

Robins die throughout the year, but breed only in spring, hence there must be more of them alive in autumn than in spring. But at both Dartington and Enniskillen there are only about the same number present in late autumn as there are in April. Hence a proportion must migrate for the winter. As regards the sex of those that migrate, at least most must be females. This is indicated by the differences in behaviour between males and females already described, and is also borne out by a study both of the times of pair-formation and of the sex ratio in autumn. These two latter points will now be discussed.

At both Dartington and Enniskillen the females which were known to be autumn residents formed pairs distinctly earlier than the apparent migrants. This would not have been expected if the latter had really shifted only locally, and strongly suggests they had farther to come than the autumn residents, *i.e.* that they were genuine migrants. Moreover, these migrants appear earlier at Dartington than at Enniskillen, which also indicates genuine migration, as Enniskillen is some 300 miles farther north. (See Table 2 opposite.)

Coming to the third point, if it is mainly the female robins which migrate, then one would expect a great preponderance of males in the autumn robin population. The latter was, in fact, demonstrated by a correspondent to the *Zoologist* as long ago as 1864, but he was so ashamed that he omitted to publish how many robins he killed to establish it. Colour-ringing gives a similar answer. Burkitt's figures showed that at Enniskillen at least three-quarters of the autumn robins were males. At Dartington the figure is not so reliable, because a number of autumn residents disappeared before their sex could be proved by observing their breeding behaviour. However, nearly all these birds of unknown sex sang well, which suggests that most were males, since of the autumn birds whose sex was known, all the cocks but only half the hens sang. Hence about three-quarters of the robins at Dartington in autumn were males. (See Table 3 on p. 106.)

To summarize this rather complicated situation: Of the female robins between one-third and one-quarter are non-migratory, and a few apparent migrants really move only locally; the rest migrate. Of the males most are non-migratory, but perhaps a small proportion of the first year males and possibly a few older males migrate.

The 'British Birds' Marking Scheme (now organized through the Bird Ringing Committee of the British Trust for Ornithology) provides a little further information on robin migration, but unfortunately, though many nestling robins have been ringed, rather few have been recovered. Of seventy nestlings ringed in England and Scotland and recovered between September and February, all except four were found within a few miles of where they were hatched. The other four were found abroad, as follows:

*Robins ringed as nestlings in Britain and recovered abroad**

Place and date of ringing		Place and date of recovery	
Staffordshire	June 1913	Gers, France	Oct. 1913
Berkshire	May 1912	Voorn, Holland	Nov. 1914
Hereford	May 1930	Seine Inférieure, France	Dec. 1933
Essex	May 1938	Basses-Pyrénées, France	Sept. 1938

The small proportion of British robins recovered abroad proves that migration occurs, but does not necessarily reflect the proportion which migrate, since it is uncertain whether dead ringed robins would be reported so readily from France as from England.

There is further evidence that British robins migrate, since

*Later records up to 1963 include three more from France (in Calvados, Vendée and the Landes) and one from Cordoba in Spain.

they have been seen in September on the island of Ushant, off Brest, and one juvenile female was collected. The British robin is also said to be a regular autumn visitor to the Channel Isles and the Normandy coast, on the latter being distinguishable from the local robins by its fuller song. The final destination of these passage migrants is not known; perhaps it is the South of France, Spain, or Portugal.

There is no evidence that robins breeding in the north of Britain winter in southern England. Thus of twenty-eight robins ringed in summer in northern England and recovered in winter, none were found in southern England, while of thirty-four ringed in winter in southern England and found in summer, none were taken in northern Britain. Also no robins ringed in Scotland or England have been found in Ireland. However, robins appear regularly in winter on islands off the Irish coast where they do not breed. For instance, a robin wintered for several years in succession on Copeland Mew Island off Belfast, and it was evidently the same individual each time, as it chose exactly the same spot for roosting. Likewise the robin is a regular winter visitor to Skokholm, off the south coast of Wales, and two birds ringed there in winter disappeared for the summer, but reappeared for each of two later winters. Such occurrences suggest that a small proportion of British robins move west and south-west in winter, but it is not, of course, known how far these island birds had travelled, nor is it quite certain that they were British robins.

Migratory robins have been seen on many small islands and lighthouses off the British coast, particularly in autumn. Seven collected at the Tuskar Rock, off south east Ireland, were all first year females, thus supporting the view that it is mainly the hens which migrate. At the Irish stations robins appear much less frequently than most song-birds, but on Fair Isle, the Isle of May, and in Norfolk, they sometimes arrive in large numbers. For instance, in the last days of September 1933, the bushes along two miles of the Norfolk coast at Blakeney held some three thousand individuals, and several were found lying on the beach too exhausted to move. These were Continental robins, which had evidently met an adverse wind after setting out on their migration. The British robin, which has been identified on passage at the Isle of May and other places, has not been seen in such large numbers.

Another large arrival of robins is mentioned by William Prynne in *A New Discovery of the Prelate's Tyranny* (1641). For his attacks on Archbishop Laud, Dr John Bastwick was in October 1637 imprisoned 'at the Islands of Scylles, when many thousands of robin redbreasts (none of which birds were seen in those islands before or since) newly arrived at the Castle there the evening before, welcomed him with their melody, and within one day or two after tooke their flight from thence, no man knows whither'.

It was at one time customary to classify birds as either residents or migrants. However, the robin comes in both categories. In South Devon all or nearly all the adult males and one-third of the females are resident, while two-thirds of the females and perhaps a few males migrate. In the most northerly parts of Britain many robins are claimed to leave in winter, suggesting that a greater proportion migrate. In parts of Germany all except some of the adult males are said to migrate, leading up to the condition in most of Scandinavia, where all migrate. The other extreme, where none migrate, holds in the Canary Islands and the Azores. The wintering grounds of the Continental robin are gradually being mapped from the recoveries of ringed birds (Fig. 4 opposite).

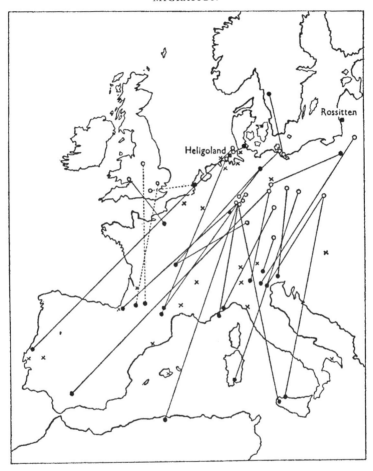

Fig. 4. Migration of robins

○ indicate places of ringing in breeding season
● indicate places of recovery in winter or on passage
✗ indicate recoveries of birds ringed as passage migrants
    at Rossitten or Heligoland

[Adapted from Drost and Schüz, *Vogelzug*; 3, p. 167, 1932.

Differences in migratory behaviour between the two sexes, such as occur in the robin, are not so uncommon as was once thought. In the Californian mocking-bird most males are resident, most females migrate, while some females stay, holding territories and singing, like resident hen robins. The males of Güldenstadt's redstart reside throughout the year in Tibet, whereas the females migrate. In chaffinch and blackbird on the continent of Europe more females than males migrate; the same is true of the song-sparrow in Ohio and of various other American birds. In all these species it is the males which tend to be less migratory than the females. Even in purely migratory species, like the British warblers, there is a sexual difference in the migratory urge, since the cocks arrive in spring some days ahead of the hens.

The tendency, possibly found in the British robin, for the first-year birds to be more migratory than the adults, has been claimed to hold for the Continental robin, and is true of song-thrush, lapwing, black-headed gull, and gannet in Britain, and of chaffinch, blackbird, tits, woodpeckers, sparrowhawk, and common buzzard in Germany. Indeed, it may well be general among species which are partial migrants.

These differences in the migration of males and females, adults and young, suggest that the migratory urge may be affected by the sexual (gonadotropic) hormones. In particular, since the male sex hormone is likely to be present in larger quantities in adult males than in either females or young males, this hormone may possibly induce birds to migrate earlier to and later from their breeding quarters. It seems significant that those hen robins which, like cocks, do not migrate in autumn, also show other characteristics of male behaviour, since they sing, fight, and maintain territories, and the same holds for those female mocking-birds which do not migrate. However, this possible effect of male sex hormone needs to be checked by experiment, and the situation is certainly complex, since in the north of Europe all the robins of both sexes migrate, while in the Canary Islands and Azores none do so.

Further, migratory Continental robins take up territories and sing in their winter quarters in Italy, Dalmatia, and North Africa. They divide the coastal plain of Algeria into individual territories just as they do in England, only with this difference, that in Algeria the birds arrive in October and leave in March. Song and fighting are at their peak at the end of October, but there is no second increase in song in the early spring, before the birds depart, which, it has been suggested, may be because most of these migrants are females. Incidentally, the wintering European robins do not appear to mix with the local Algerian robins, which are found in the hills at higher altitudes, chiefly above 2,000 feet.

The robin does not altogether lose its aggressiveness even on migration. Some seen arriving in groups of twelve to twenty at Salerno in mid-March settled in the bushes behind the beach and immediately started chasing and scolding, though there was no singing. More remarkable is the following incident described in 1833: 'On the voyage from London northward, on the 16 September, when off the coast of Yorkshire … several small birds alighted on the vessel… On the following day other species made their appearance … all … left the vessel on the first night after their appearance, except two robins, which remained for some time, and which, with the characteristic effrontry of their species, stationed themselves, the one on the front of the vessel and the other at the stem, and fought at the least intrusion into each other's territory.'

The migratory urge of birds was first studied experimentally by W. Rowan in Canada. He considered that it must depend on the state of the sex organs, and showed that, by providing artificial daylight to caged American juncos, their sex organs could be brought to maturity in mid-winter. Rowan's further experiments to show that such birds would, if released, migrate, were inconclusive, but this result has since been obtained by a Californian worker. Several years earlier, ingenious experiments to the same effect were carried out on migratory Continental robins trapped on Heligoland. It had previously been found that, if migratory birds are trapped and kept in cages, they become extremely restless at night during the migratory seasons. When six captive robins were given extra light from sunset until 1 a.m. during the winter, their sex organs matured earlier, and in addition their migratory restlessness started earlier, than in six other captive robins kept as controls which experienced a normal daylength. In another six robins, which were kept indoors and subjected to a constant eight-hour day, the sex organs did not develop and migratory restlessness did not appear.

These experiments suggest that the same hormones which stimulate the sex organs also stimulate spring migratory behaviour. But the immediate stimulus to migrate is different, and is correlated with temperature, wind, and time of day. During the spring migratory period, restlessness occurs mainly on warm fine nights, and it is inhibited on cold and stormy nights, and during the daytime. For instance, at the end of March, migratory restlessness was at once shown by some captive robins when the room temperature rose from 5∞C to 20∞C. In autumn, on the other hand, migratory restlessness is stimulated by a sudden fall in temperature. Experiments have not yet been carried out on the relation between daylength, the gonadotropic hormones, and the onset of autumn migration.

To some of the Greek coasts the robin is purely a winter visitor. The common redstart, on the other hand, is purely a summer visitor. Perhaps this was why Aristotle considered that these two species changed into each other. Pliny followed him: 'The bird which is named Erithacus (i. robin, or redbreast) in winter; the same is Phœnicurus (i. redtaile) all summer long.' It was William Turner who refuted this view, in his commentary on the birds of Aristotle and Pliny written in 1544, but the problem was easier for Turner since both species nested where he lived.

Aristotle, of course, knew of the existence of bird migration, and described it in a number of species, though, as in the above case, some of the seasonal appearances and disappearances were attributed to other causes. 'Swallows, for instance, have often been found in holes, quite denuded of feathers.' But a highly distinguished company, including Linnæus, Gilbert White, Sigismond King of Poland, and Dr Johnson, were in error over the swallow, Johnson remarking: 'Swallows certainly sleep all the winter. A number of them conglobulate together, by flying round and round, and then all in a heap throw themselves under water and lie in the bed of a river.'*

This chapter may be concluded with a note on the distances moved by those British robins which are non-migratory. Seven cocks and one hen resident at Dartington shifted their territories for distances of up to a few hundred yards. Any which may have shifted farther than this would probably not be noticed again. Other information is provided by the returns of the British Birds Marking Scheme. Of 111 British

---

* But the poor-will, an American nightjar, has now been found to hibernate in rock crevices in California!

robins trapped as residents which survived at least a year following their capture, 105 were found dead within a mile of where ringed, only six farther away, showing that extremely few British robins shift their ground far, once they have taken up territory for the first time. Of the six robins which shifted, three travelled under five miles, the other three, together with two others found within a year of ringing (and so excluded from the above analysis), moved as follows:

British robins trapped as residents which have moved over 10 miles

Place and date of ringing		Place and date of recovery		Distance in miles
Whitchurch, Hants.	April 1925	Basingstoke, Hants.	May 1927	11
Malvern, Worcs.	July 1938	Mitcheldean, Glos.	April 1939	19
Hilton, Hunts.	Nov. 1938	Barkway, Herts.	Dec. 1939	20
Isle of May	March 1936	Duns, Berwick	Feb. 1938	29
Kilnsea, Yorks.	Oct. 1938	Thomer, Yorks.	June 1939	66

A much greater proportion of the juvenile than of the adult robins shift, as is the case in most song-birds, but extremely few move more than a few miles. Of ninety ringed nestlings which survived at least until their first September and were later found dead in Britain, forty, or nearly half, were found dead at least a mile away from where ringed. But of these, as many as twenty-nine were found within five miles of where ringed, six more between six and eight miles, and only five others rather farther, as follows:

British robins ringed as nestlings which moved over 10 miles

Place and date of ringing		Place and date of recovery		Distance in miles
Broughty Ferry, Dundee	May 1922	Gilston, Fife	July 1923	14
Ullswater, Cumb.	May 1926	Langwathby, Cumb.	Dec. 1926	17
Bridge of Earn, Perth	July 1923	Wormit, Fife	Oct. 1923	17
Banbury, Oxford	May 1936	Hillmorton, Rugby	Spring 1939	22
Abernethy, Perth	May 1925	Arbroath, Forfar	Nov. 1925	32

To complete the record of the movements of nestlings, the four already noted as recovered abroad should be added. The individual which went to Holland should probably be classified as shifting or lost rather than as migrating, as this is a curious direction for a migrant.*

In the three summers of 1935-7, 121 nestling robins were ringed near Dartington. Of these birds thirteen were seen again after the autumn moult, of which seven, and almost certainly two more, were males and held territories, and one, and probably one other, were females. The sex of the other two was not determined. One male occupied the territory of its former parents, another held an adjacent territory, the others were varying distances up to 900 yards from where ringed. If any settled farther away than this, they would probably not have been noticed. Similarly, Mrs Nice found 13 per cent of the song-sparrows ringed as nestlings returned in the following spring and, as in the robin, many more males than females returned.

These results show what short distances most robins travel during their lives. Indeed if, as in one case at Dartington, a young bird occupies the territory of its former parents, it is possible for it to live to old age without ever moving outside a circle of some 300 yards in radius, a distance which it could easily fly in less than a minute.

The available evidence suggests that each robin population is closely adapted to local conditions. As regards the proportion of individuals which migrate, the date at which the migrants return, the date at which breeding starts and finishes,

---

* By 1963 there had been ten further records of British robins ringed as nestlings moving over 10 miles within Britain, the furthest distance being 130 miles, the rest less than 60 miles. Several others ringed as adults also moved more than 10 miles.

and the number of eggs in the clutch, there are definite average differences between the robins of different places. Such differences are likely to arise only in a species in which most individuals breed close to where they are born.

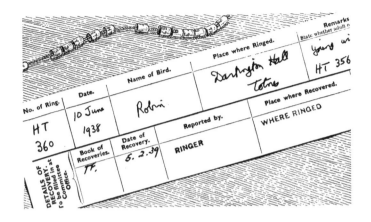

9

Age

> Touching the Length and Shortness of Life in Living
> Creatures, the information which may be had is but
> slender, observation is Negligent and Tradition Fab-
> ulous. In Tame Creatures their Degenerate Life cor-
> rupteth them; in wild Creatures their exposing to all
> weathers often intercepteth them.
>
> FRANCIS BACON* *The History Naturall and
> Experimentall of Life and Death* (1638)

In July 1938 Burkitt revisited Enniskillen and saw a hen robin
still alive and bearing rings placed on her in December 1927.
She must therefore have been at least eleven years old, and is
the oldest robin so far recorded. A Continental robin ringed
on Heligoland has been recovered in its eleventh year, and the

---

*A translation by William Rawley of the *Historia Vitæ et Mortis*, 1622–3.

oldest yet found under the British Birds Marking Scheme died in its ninth year.

Until recently the only information on the length of life in birds came from those kept in captivity, but ringing returns are now beginning to provide data on the age to which birds live in a wild state. The records from both sources show that many small song-birds can, like the robin, live for ten or eleven years, in some cases up to fifteen years, while in rare cases an age of twenty years has been recorded. Larger birds can live longer, and captive parrots, eagles, and swans have been credited with living over a hundred years. Indeed an even greater age has sometimes been claimed, but on very dubious evidence.

The life-span of a robin or other small bird is about four times as long as that of a small mammal of similar size, such as a mouse, and is of about the same length as that of a dog or cat. Also, while a robin's life-span is only about a ninth of that of a man, all the bird's activities are much more rapid, so that its time-scale is different. The pulse-rate of a redstart is 980 beats per minute, which is about fourteen times as fast as that of a man, and, since the robin is closely related to the redstart and of about the same size, its pulse-rate is probably similar. Reckoned in terms of heart beats, an eleven-year-old robin is equivalent to a man 150 years old.

'Of all sanguineous and hot Animals Birds are the longest lived,' Ray wrote in his *Ornithology* in 1678 and, after recording that a caged linnet lived for fourteen years and a goldfinch for twenty, concluded: 'And there is no doubt but birds that enjoy their liberty, living at large in the open air, and using their natural and proper food, in gathering of which they also exercise their bodies, live much longer than those that are imprisoned in houses and cages.' The latter incautious guess is readily disproved by a simple calculation made by Alfred Russell Wallace in 1889. Assuming that each robin lives for ten years and that each pair produces ten eggs a year, then, if they and their offspring are unmolested, a single pair will in ten years have multiplied to more than twenty millions. As there is no reason to think that the robin population is increasing at all, it is clear that an exceedingly large number of robins must die or be killed each year. Yet, as Wallace adds, 'We see nothing, or almost nothing, of this tremendous slaughter of the innocents going on all around us.'

Clearly the average length of life of a wild robin must be very different from the age to which it can survive under favourable circumstances. The first estimate for the average age of any wild bird was that made for the robin by Burkitt. Burkitt assumed that the number of new robins produced each year must be balanced by a corresponding number of deaths among the existing adults, and, after assessing the replacement rate, he calculated the average age of a robin to be about two years and ten months. Low though this figure seems, it is almost certainly much too high.

Another method is now available for estimating the average age of the robin, by examining the length of life of those robins ringed under the British Birds Marking Scheme and later found dead. The age of a ringed bird at its death is, of course, known by looking up the ring-number in the files to see when the ring was put on. Only a very small proportion of the individuals that are ringed are later recovered dead, just under 1 per cent of those ringed as nestlings, and just under 3½ per cent of those marked as adults.* Further, as can be seen

---

* Between 1931 and 1938, 5,115 robins were ringed as nestlings, and 5,528 were trapped and marked as adults. Of these, 50 of the nestlings and 191 of the trapped adults were later found dead.

from the causes of death listed in the next chapter, the recovered robins represent only those killed in particular ways. Hence all that the average age of the recovered birds certainly shows is the length of life of robins found dead by human beings. It is somewhat like calculating the average age of English people from the records of deaths in motor accidents, or from the gravestone inscriptions in a particular cemetery. The sample is a small one, but, in the case of the ringed robins, there is no reason to think that it is atypical, as regards age, of the population as a whole.

It is convenient to calculate robin survival in terms of a year which starts on 1 August and ends on 31 July. The calendar year, from 1 January to 31 December, is rather inconvenient, as young robins are born only in summer. On 1 August, the robin population consists of a number of adults of various ages which have just finished breeding, and a number of juveniles which were hatched during the previous three months. An analysis of the ringing recoveries of trapped adults, of unknown age when caught, shows that of those alive on 1 August as many as 62 per cent, or well over half, die before a further year has passed. Indeed, their average survival-time works out at the astonishingly low figure of thirteen months, less than half what Burkitt estimated it to be. Of the juveniles alive on 1 August, an even higher proportion, 72 per cent or nearly three-quarters, die before a year has elapsed, and their survival-time is only about twelve months. (See table opposite.)

A similar analysis of the ringing recoveries of other British birds shows that most of them can expect to live, on the average, for rather longer than a robin, a song-thrush or starling for about a year and a half, a blackbird for nearly two years, a woodcock for just over two years, and a lapwing or black-

Survival of ringed Robins

| | ADULTS | | JUVENILES | |
	Alive on 1st Aug	Dying during year	Alive on 1st Aug	Dying during year
1st year	129	80	130	94
2nd year	49	29	36	17
3rd year	20	12	19	14
4th year	8	6	5	3
5th year	2	2	2	1
9th year	—	—	1	1

	ADULT	JUVENILES
Percentage dying during first year	62%	72%
Average age at death (or expectation of further life on 1 Aug.	13.3 months	12.1 months

Table based on the subsequent recoveries of (i) adults ringed before summer 1938, and (ii) young ringed before 1939, which survived at least until 1 August following ringing.

headed gull for two and a half years. These, of course, are average figures. Many individuals live for a shorter time than this, and a few for much longer, up to ten years, fifteen years or even more. Again, in the American song-sparrow, Mrs Nice found that about 40 per cent of her colour-ringed cock birds disappeared each year, which means that their average length of life was about two and a half years. Similarly a Dutch worker found that about 60 per cent of his colour-ringed redstarts disappeared each year, which gives them an average survival-time of about a year, a figure closely similar to that obtained above for the robin, to which the redstart is allied. All this work is, therefore, to the same effect, that wild birds live on the average for a far shorter length of time than popularly believed; and the robin has the shortest life of any.

The calculations for the robin were started from 1 August, rather than earlier in the summer, in order to exclude the period in which the fledgling bird is dependent on its parents, and when it is particularly vulnerable. But the figures suggest that, even when the first month or two of juvenile life is excluded, first-year robins survive less well than older birds. Of the older birds 62 per cent, and of the juveniles 72 per cent, die within a year from 1 August. The totals available for analysis are too small for this difference to be entirely established, but strong corroboration is afforded by the existence of a similar difference in other British birds investigated. It is probably due to the greater experience of the older birds in avoiding dangers. Experience is acquired by the juveniles only at a price.

In some other species, such as the blackbird and lapwing, there are sufficient ringing returns of older individuals to indicate that the expectation of further life is about the same for a two-year-old, three-year-old, four-year-old bird, and so on, up to the age at which too few further individuals are left for the figures at present available to be reliable. The figures for the robin are not inconsistent with a similar conclusion but are too few for this to be established. The returns for these other species also show that under natural conditions extremely few wild birds reach 'old age', *i.e.* an age approaching that to which they can survive under favourable circumstances.

This state of affairs is in marked contrast to the survival of civilized human population. For example, in 1930 the expectation of further life for a four-year-old male inhabitant of the United States was another sixty years, as compared with a possible human life-span of about a hundred years. In the robin the expectation of further life for a two-month-old bird is just under another year, as compared with a possible life-span of at least eleven years. Hence the human juvenile can expect to live for about six-tenths, but the juvenile robin for less than one-tenth, of its potential life-span. Among men, elderly individuals, *i.e.* those near the end of the potential life-span, are common, among wild birds they are rare. Also, omitting the first three years of life, the expectation of further life for a man decreases markedly and steadily with increasing age, whereas in a wild bird it does not appreciably decrease with increasing age up to the age by which nearly all individuals have died.

The survival of some of the populations of rats, mice, or insects raised in laboratories follows a course very similar to that of a modern human population. Their expectation of life is a high percentage of their potential life-span, elderly individuals are relatively common in a stationary population, and their expectation of further life decreases markedly with increasing age. On the other hand, burial records indicate that the survival of the people living in Ancient Rome was of a type more similar to that of wild birds. As compared with a modern civilized population, their expectation of life was a relatively low percentage of the potential life-span, elderly persons were relatively much less common, and the expectation of further life decreased little with increasing age during a long period between adolescence and late middle age. While much more information is needed, it may be guessed that the type of survival shown by robins and the people of Ancient Rome will be found to apply to most wild birds, in which 'their exposing to all weathers often intercepteth them'. Probably only protected animals, such as civilized man and laboratory rats, have the type of life-curve on which the estimates of life insurance companies are based. These conclusions are shown diagrammatically in Fig. 5 overleaf.

As discussed in Chapter 5, about one-fifth of the cock

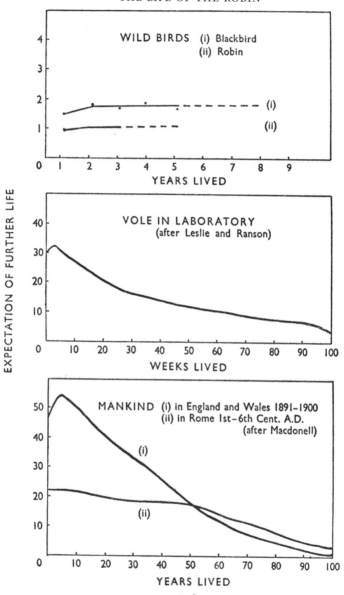

WILD BIRDS (i) Blackbird
(ii) Robin

(i)

(ii)

YEARS LIVED

VOLE IN LABORATORY
(after Leslie and Ranson)

WEEKS LIVED

MANKIND (i) in England and Wales 1891-1900
(ii) in Rome 1st-6th Cent. A.D.
(after Macdonell)

(i)

(ii)

YEARS LIVED

EXPECTATION OF FURTHER LIFE

robins fail to get mates. No corresponding unmated hens were observed. Hence if, as one might expect, the sex ratio among nestling robins is about equal, hen robins would seem to have a higher mortality than cocks. Unfortunately this cannot be checked from the ringing returns as the sex of the birds was not usually recorded.

A rough check is available to test whether the figures based on the ringing recoveries give a survival-rate for the robin which is at all probable. There is no reason to think that the robin population of Britain is either increasing or decreasing. This cannot be proved, but so widespread is the interest in birds that it is almost certain that any marked change in one direction or the other would have been recorded. If the robin population is roughly stationary, then the adults dying during the year must be replaced by an approximately equal number of young which survive to breed. One can, in fact, draw up a balance-sheet. The figures already given suggest that about 62 per cent of the adult robins die during the year. Hence every hundred adult robins must each year give rise to about sixty-two young which survive to breed. Is this likely from what is known of the survival of young robins?

In the previous calculations, the survival of juvenile robins from the time of leaving the nest until the following 1 August was omitted. Actually, of all those robins ringed as nestlings and later recovered, as many as 26 per cent, or a quarter of the total, were found dead within one to three months, before reaching the first August of their lives; this figure may somewhat exaggerate the true mortality, as some bird-ringers have made a special search near the nests for dead ringed birds in this period. Bearing in mind a possible bias due to this cause,

Opposite: Fig. 5. Expectation of Life

of all those robins ringed as nestlings and later recovered, as many as 77 per cent were found dead between the time of leaving the nest and the end of their first year of life; only about 23 per cent survived until the time of breeding.*

Since sixty-two out of every hundred adult robins die each year, sixty-two new breeding individuals are needed to replace them. The calculation in the previous paragraph suggests that, of every hundred young robins leaving the nest, only twenty-three survive to the following breeding season. Hence to get sixty-two new breeding individuals, about 270 fledglings are required. Therefore, to balance the annual mortality, every hundred adults must each year produce 270 fledglings. As about one-fifth of the cock robins are unmated, a population of a hundred adult robins consists of forty-five breeding pairs and ten unmated males. Hence the 270 fledglings have to be produced by forty-five pairs, an average of six fledglings per pair. Is that reasonable?

The robin lays on the average five eggs in each clutch and is commonly double-brooded. In 1945, I published a request to ornithologists to send me all their nest records for the robin. When the resulting information was summarized, it was found that, on the average, 55 per cent of the eggs in completed clutches gave rise to young which successfully left the nest.†

---

*Based on 144 recoveries, of which 111 occurred between the time of leaving the nest and the following 1 June. This calculation is similar to that on p. 122 except that every recovery was taken into consideration, whereas on p. 122 those individuals recovered before their first August of life were omitted. As mentioned above, the true mortality in the first year is perhaps rather lower than 77 per cent.

† Of 1,426 robins' eggs studied, 71 per cent hatched successfully; of 1,866 newly hatched young studied, 77 per cent left the nest successfully. 77 per cent of 71 gives 55 per cent.

Hence, if the robin is double-brooded, each pair lays ten eggs a year, from which 55 per cent, or nearly six, young successfully leave the nest. As six young per pair per year is the number required, the apparent balance between robin births and robin deaths is as close as could be expected from these rather rough figures. The ringing returns evidently give an approximately correct picture of the mortality.

The only way to obtain a true estimate of the average length of life of wild robins will be to make an accurate survey of a large number of ringed robins over a period of years and over a wide area. Unfortunately, while the Dartington area was well suited for studying the behaviour of robins, it was not large enough for studying their survival, as only a small shift of territory would take a bird outside the area, where it was easy to overlook it. Indeed, several individuals which I thought had died were really inhabiting adjoining ground, and I came to know of their existence only through the reports of friends, who had seen them in their gardens. On the average, 80 per cent of the cock robins resident each breeding season at Dartington had disappeared by the following breeding season; as some of these birds had certainly shifted and not died, the result is not inconsistent with a 60 per cent annual mortality.

Until a large-scale survey is undertaken over a wide area, the figures for robin-survival based on ringing returns are the best available, but it should be remembered that, though they probably apply to the robin population as a whole, they certainly apply only to the robins found dead by human beings. Those who doubt the present evidence, because they cannot believe that a robin should have so short an average life as one year, must bear in mind that, if the average life were appreciably longer and the replacement rate remained the same,

Britain's robins would be increasing so rapidly as to become a plague.

Postscript

Through the courtesy of the British Trust for Ornithology, I in January 1965 examined the far greater number of recoveries of British robins ringed as young up to the end of 1960 and later found dead, by then some 450 in all. These new data do not change the argument developed in the present chapter, so I left the earlier figures unaltered. Combining the newer figures with the earlier ones, 29 per cent of the ringed young were found dead before their first August 1st (a slightly higher percentage than before), while of those which survived at least until their first August 1st, 69 per cent were recovered within the next year, and an average of 57 per cent in each subsequent year, both figures being slightly lower than those based on the earlier sample. This gives an adult robin an expectation of further life of 1¼ years, which should be regarded as a more accurate figure than that of 13 months given earlier.

10

Food, Feeding, and Being Fed Upon

Then, as they were coming in from abroad, they es-
pied a little robin with a great spider in his mouth; so
the Interpreter said, 'Look here'. So they looked, and
Mercy wondered; but Christiana said, 'What a dispar-
agement is it to such a pretty little bird as the robin-
redbreast is, he being also a bird above many, that
loveth to maintain a kind of sociableness with man; I
had thought they had lived on crumbs of bread, or
upon such other harmless matter. I like him worse
than I did.'

JOHN BUNYAN: *Pilgrim's Progress* (1678)

The above apparently has the distinction of being the only
disparaging reference to the robin in the whole of literature,
the Interpreter continuing: 'This robin is an emblem very apt

to set forth some Professors by; for to sight they are as this robin, pretty of Note, Colour, and Carriage, they seem also to have a very great Love for Professors that are sincere... but when they are by themselves, *as the robin,* they can catch and gobble up *spiders*, they can change their Diet, drink *Iniquity*, and swallow down *Sin* like water.'

That the robin takes animal food has been on record since Aristotle mentioned that it fed on worms. In summer it feeds particularly on small caterpillars and in winter particularly on small beetles, but its diet also includes a great variety of other insects, both adults and larvæ, including those of blow-flies, crane-flies, plant bugs, parasitic flies, gall insects, earwigs, and ants. In fact it takes almost any insect that it can get, except that it normally rejects a distasteful butterfly. Wordsworth's distress that a robin should chase a butterfly was unnecessary, and later more accurate observers have found that a robin, after catching a butterfly, releases it without trying to recover it. The robin also eats spiders, centipedes, earthworms, and small land molluscs, while vegetable food includes small seeds and many small fruits, including currant, raspberry, blackberry, whortleberry, ivy, hawthorn, and yew.

Except for eating a few garden fruits, the robin does no direct harm to man, though its destruction of spiders, parasitic flies, and earthworms might also be reckoned against it. On the other hand, its food includes many injurious insects and a small amount of weed seeds. Collinge analysed the food in fourteen robin stomachs as '43.5 per cent beneficial, 48.5 per cent neutral, and 8 per cent injurious' to man. But this classification is not justified. Even should 43.5 per cent of the robin's food consist of harmful insects, it does not follow that the bird is to that extent beneficial. The harmful forms include many caterpillars, whose numbers may well be controlled primarily by parasitic ichneumon flies. If the robin takes parasitized and unparasitized caterpillars alike, then it may hinder the controlling influence of the parasites, and its long-term influence on the caterpillar population cannot be determined without a detailed and quantitative investigation. At present, far too little is known to say whether or not the robin is economically beneficial or harmful to man, and the same applies to all other insect-eating birds.

The robin starts to feed earlier in the morning, and stops later in the evening, than most other song-birds, and its large eye perhaps helps it to see in dim light. It takes most of its food off the ground, searching over the soil and in litter. Characteristically, it uses the lower branches of trees or bushes as look-out posts, from which it flies to the ground on detecting an insect. The bird's habit of waiting on the gardener is well known, and it hops down to pick up the worms and insect larvæ turned up by the spade. It may similarly attend on moles working. It also finds a little of its insect food on the branches and leaves, and occasionally catches insects on the wing. An unusual record is of one diving into the water of a garden pond and taking out a small roach, and there are several instances of it taking something off the surface of a stream.

After feeding on ivy berries, a robin often ejects the seeds from its mouth. Not infrequently, definite pellets are brought up. This was common with the aviary birds. One pellet from a wild robin contained twenty raspberry seeds, two white and two red currants, the remains of flies and an earwig. Another ejected the chitinous remains of various insects, while a third, which spent the winter inside a church, produced a pellet consisting chiefly of the remains of bluebottle flies. A fourth pellet contained unidentifiable fruit remains and a small piece of brick – like many other birds, the robin swallows grit to help grind up the food in the gizzard. The habit of ejecting indigestible food-remains from the mouth in the form of pellets occurs in many other song-birds.

Insects provide the bulk of the robin's food throughout the year, but in winter they are not so easily found. At the latter season, many of the hen robins have migrated south to a warmer climate, where insects are presumably more abundant, while the resident robins become bolder, feeding more in the open, and regularly coming to fowl runs, scrap heaps, back doors, bird tables and other human sources of food, even visiting inside the house when encouraged. Fat seems in particular demand, and several tamers of robins have noted the bird's fondness for butter. Three tame robins which regularly took butter would not take margarine, though a fourth individual did so freely. In hard weather the robin has also been reported feeding on the carcasses of small birds, and on the fat of meat hanging in butchers' shops. Indeed, the bird will sample almost anything, one house-tame bird trying both floor polish and sulphate of potash used for spraying plants.

At irregular intervals Britain experiences an unusually severe winter, and from the year 1408 onwards, scattered records are available for the heavy mortality then produced among wintering birds. For the hard winters of the twentieth century, which occurred during the early months of 1917, 1929, 1940, 1941, 1945, and 1947, detailed records are available through correspondents to the magazine *British Birds*. Various of these correspondents put the robin mortality in such winters at between 60 and 80 per cent, and similar estimates are given for the dunnock, wren, great, blue, coal, and marsh tits. Other insectivorous birds, notably the tree creeper, long-tailed tit, bearded tit, goldcrest, Dartford warbler, and stonechat, suffer much more severely than the robin, some-times being brought temporarily near to extinction. Thrushes, particularly the redwing, have also been found dead in thou-sands. On the other hand, most finches and the members of the crow family suffer much less than the robin.

It is the length rather than the intensity of a cold spell which matters. The robin lays up a store of fat for the winter, and this probably helps it to tide over short periods with little or no food. Data from the Reverend John Lees and Colonel Meinertzhagen show that in summer the cock British robin often weighs only 18 grams, whereas in winter it commonly weighs 22 grams, and sometimes up to 25 grams. French robins show a similar increase in weight in winter, and parallel variations are found in many other song-birds.

In the hard winter of 1880-81, five robins came into a Devon house and continued there, feeding on scraps, until the weather moderated a fortnight later. Likewise in the cold spell of 1940, an Oxford ornithologist had as many as eleven robins visiting his bird-trap in two days, sometimes four individuals fed there together, and though the cock owning the territory occasionally chased the intruders, it often did not do so. He therefore supposed that territorial behaviour may be aban-doned in cold weather, but I doubt if this is correct, because six individuals that I watched during the exceptionally pro-longed and severe cold weather of 1947 often trespassed to feed at a common source of food, but each retained its aggres-siveness. The same was found by another observer who reg-ularly fed several robins during this period, and who found that fights were frequent whenever one individual tried to take food from him in the territory of another.

Correlated with its dependence on human food in winter, the robin becomes very confiding, a habit which is sometimes its undoing, for it falls a frequent prey to cats. 'These villain-ous false cats were made for mice and rats and not for birdes small.' A friend informs me that when her cat ate a robin it afterwards vomited it up and thereafter would not eat robins, though it still caught them. Similarly a correspondent to the *Zoologist* in 1863 stated that he never knew of a cat which would eat a robin, and another correspondent wrote that his tame fox was always sick if given a robin. Likewise there is a country belief, reported to *Notes and Queries* in 1850, that weasel and wild-cat will not molest or eat a robin. Usually animals learn to avoid those poisonous creatures, such as cer-tain snakes or caterpillars, which show bright warning colours, but since cats continue to catch robins, it seems that the robin's appearance is not sufficiently striking for the cat to recognize it until it has actually caught it – unless, indeed, cats catch robins just for sport.

The records of robins found dead under the British Birds Marking Scheme state the cause of death where known. This, of course, gives only a very incomplete picture of the ways in which robins die since, as already mentioned, only 1 per cent of fledgling robins, and 3½ per cent of wintering adult robins, are later found dead by human beings. The rest leave no trace. Further, of those which have been reported, well over two thirds were simply 'found dead', and the cause of death was not known. Who killed cock robin is, for the most part, still a mystery. Of 110 cases in which the cause of death was reported, cats accounted for forty-four, mouse-traps for twenty-four, other traps for four, eleven were found dead on roads, presumably killed by cars or through hitting the telegraph wires, ten were found drowned, four were killed by other robins, and two by tawny owls, and the following accounted for one death each: stoat, weasel, rat, dog, sparrowhawk, little owl, 'owl', great grey shrike, frozen, caught in netting, hit wire netting in fog.

The large number caught in mouse-traps is correlated with the robin's intensive search for food. The number found drowned was a surprise, and difficult to account for unless the bird's habit of plunging into water after food, mentioned earlier, is commoner than supposed and likely to lead to disaster.

There are occasional records of more unusual ways of death. Thus a parent and a young robin were found dead, each with a portion of the same horsehair fixed firmly in the gizzard. Another robin was found dead with its neck caught between two upright slates forming part of a garden fence. A remarkable accident occurred to another robin which, presumably when preening itself, managed to embed the lower half of its beak deep into the skin of its neck, and all efforts to disentangle itself failed. This bird was fortunately found by an ornithologist who, by almost strangling the bird, was able to disengage its beak, and the robin soon recovered and flew off.

Mankind has also to be included as an enemy of the robin. In his great zoological text-book Baron Cuvier did not omit to note that 'In France, the redbreasts are more numerous in Lorraine and Burgundy than elsewhere. They are very much sought after there, and their flesh acquires an excellent fat, which renders it a very delicate meat.' Again, Charles Waterton wrote: 'At the bird-market near the rotunda in Rome, I have counted more than fifty robin-redbreasts lying dead on one stall. "Is it possible," said I to the vendor, "that you can kill and eat these pretty songsters?" "Yes," said he with a grin, "and if you will take a dozen of them home for your dinner today, you will come back for two dozen tomorrow."'

Gurney saw hundreds of robins in the Algiers poultry market in February, and Godard, writing in 1916, states that in the district of Le Var, round Toulon, the robins destroyed in one season totalled 20,000.

In the seventeenth century the robin was eaten in England, as noted in the quotation from Thomas Muffett in Chapter 1. Again, William Salmon in his *Seblasium, the Compleat English Physician or the Druggist's Shop Opened*, writes of the robin in 1693, 'It is good Food, and its Medecinal Virtues are the same with those of the Sparrow, to which I refer you.' Turning to the sparrow, one finds, 'The Flesh of the Hedg Sparrow, Authors say, is admirable to break the Stone and expell it, being broiled and eaten with Salt, or gently calcin'd, or burnt, dried to Ashes and so taken R_x of the said Ashes or Pouder ʒi, Pouder of Winter Cherries ʒi, Salt of Tartar, Sal prunellae aa gr.x mix for a Dose for the same purpose, and to open all Obstructions of the Reins, Ureters, and Bladder.' The same ashes or pouder mixed with other ingredients are prescribed for the epilepsie and other distempers of the brain, while the dung with other ingredients was used 'to loosen Children's Bellies and to carry off acrophilous Humors', and also cleansed the skin of scurf, sunburning, and freckles.

By the nineteenth century, though still esteemed a delicacy on the Continent, the robin seems to have died out as an English food and as a medicine, though the Victorians sometimes killed robins to use their feathers on Christmas cards, and also as trimmings for ladies' dresses. In the twentieth century the robin does not seem to be killed by anyone in Britain except an occasional boy with a catapult. Yet in Britain, as noted in the last chapter, about 60 per cent of all adult robins die each year. Indeed, they must do so, since the population stays roughly constant. How most of these robins die is not known, but the problem is of such magnitude that it invites speculation. The death of cock robin is not a case of individual murder but of mass destruction.

The four most likely causes of heavy mortality are old age, disease, animals of prey, and starvation. The first of these has already been ruled out, since as shown in the last chapter extremely few robins reach old age. Little is known about disease in song-birds. Serious disease has not been reported in robins, and though it could easily have been overlooked, a widespread epidemic would be unlikely to have escaped attention completely. Also, the robin lives solitarily, so that contagious disease has less chance of spreading than in the case of a species which lives in flocks.

Of the birds of prey, several species have been recorded as taking an occasional robin, but only the sparrowhawk is at all a serious enemy. The sparrowhawk eats many other kinds of birds in greater numbers than the robin, and as gamekeepers keep down the numbers of the sparrowhawk, its influence on the robin population is probably small. Cats perhaps catch many robins in built-up areas, but over most of the countryside the deaths due to cats must be few. Likewise stoats, rats, and other ground vermin may eat many young robins, but they cannot account for most of the 60 per cent adult mortality.

By elimination, food shortage seems the only likely cause of widespread death among adult robins. Obviously food is not limiting throughout the year. For instance, it is in excess in late spring, since the adult robin can then find enough food not only for itself, but also for its brood. It is perhaps in midwinter that food is short. This is certainly so in severe winters, while the assiduity with which wintering robins seek for household scraps, and the fact that many hens migrate, suggest that natural food may be difficult to find even in a normal winter. Both at Dartington and in Ireland, quite a number of the ringed robins disappeared in winter. However, many others disappeared in early autumn, after breeding, and before October, which might suggest that the moult or summer drought is a time of danger. But the figures are too few for definite conclusions, and some of the disappearances may have been due to shifting and not to death.

In early April 1945, I returned for a few days to Dartington and took a breeding census of the robins in the area first studied ten years before. As shown in Fig. 6 (p. 145) the population was still at about the same figure. Such stability of population does not attract attention, but it is, none the less, remarkable. Clearly something must keep the robin drastically in check. A similar situation has been shown to hold for the heron, in which the population decreases during and recovers after the occasional cold winters, but, what is much more noteworthy, it stays at about the same density for all the intervening years. The recovery after a cold winter shows that the heron is capable of rapid increase, yet the numbers do not rise during a succession of normal seasons.

Some have claimed that in birds clutch-size is adjusted to prevent over-population. But as discussed in Chapter 7, it is much more likely that clutch-size is adapted to the maximum number of young which the parents can safely rear. Further, if clutch-size remains approximately constant from year to year, as it does in the robin and many other birds, it cannot regulate population density. It could do this only if the number of eggs laid was larger when the bird was scarce and smaller when it was abundant, which is not the case.

To keep such birds as the robin and the heron in check, there must be one or more mortality factors, the incidence of which varies with the density of the bird, so that in years when the birds are abundant a greater proportion, and in years when they are scarce a smaller proportion, are removed. Only in this way can a population be prevented from catastrophic increase or decrease. Disease, animals of prey, and food supply are all factors capable of controlling a population in this way, but too little is yet known of the economics of the robin and other birds to say which of them may be important. Another claim of a quite different nature has also been made, namely that robins and other birds control their own population density, through their territorial behaviour; but this merits detailed examination in a chapter to itself.

I I

The Significance of Territory

*Unicum arbustum haud alit duos erithacos.**

ZENODOTUS (third century B.C.)

The song, fighting, and display of the robin all centre round the acquiring and maintenance of a territory. It should be stressed that if this fighting is to result in a territory the retreat of the intruding robin is quite as important a part of the behaviour as attack by the owner. The robin not infrequently attacks dunnocks, but the latter merely evade and return, so that the robin does not achieve any isolation by its attacks on these birds.

The territory is a song area, and many robin territories have natural boundaries where song-posts stop, as, for

---

*One bush does not shelter two robins.

instance, at the edge of a wood bordering a field. Indeed the size of one territory was increased by the erection of some scaffolding at the edge of a wood, for the bird then sang regularly from the scaffolding. Where natural boundaries do not exist, the territorial limits are still fairly sharply defined, and are determined by encounters with neighbours. As at Munich in 1938, such encounters are primarily vocal, and may similarly lead to a shift in the territorial boundary.

Each year the breeding and winter territories of the robins in part of the Dartington area were mapped, as shown in Fig. 6 on pages 144-5 overleaf. The territories were usually shaped fairly evenly, but in April 1936 one pair (M2, F) had to cross sixty yards of open ground to get from one part of their territory to the other, as did a similarly situated pair in the following year, while in the next year an unmated male (M18) had an even longer distance to cross; but he, unlike the previous owners, used the intervening buildings as song-posts. Burkitt found a robin whose territory consisted of three parallel hedges in a field, and here also the bird had to cross open ground to get from one part to the next.

The quotation at the head of this chapter shows that the solitary nature of the robin has been recognized for a long time. Gesner (1555) noted 'Erithacus avis est solitaria', and Olina (1622) wrote 'Ha per proprio dove stanza di non comportavi compagno, perseguitando con ogni sforzo, chi gli sturba il suo possesso'. (It has a peculiarity that it cannot abide a companion in the place where it lives and will attack with all its strength any who dispute this claim.) Buffon (1771-83) also mentions that the male robin 'chases all the birds of his own species, and drives them from his little settlement'.

The robin spends so much of its life acquiring and maintaining a territory, and the territory is so definite, that it is

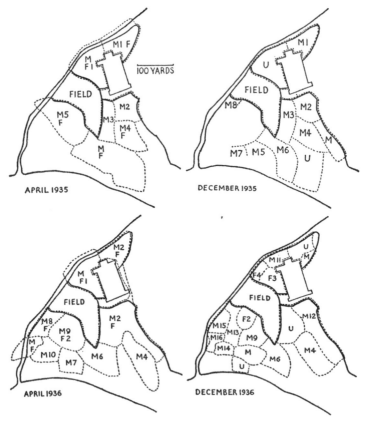

Fig. 6. Robin territories at Dartington

Boundaries of robin territories are indicated by dotted lines. The occupied ground is woodland, orchards, and quarries. The unoccupied ground is mostly fields, with a large building near top of map.

M= territory of male   M ⎫
                             F ⎭ = territory of pair

F= territory of female   U= territory of bird of unknown sex

Individual robins which stayed long enough to appear on more than one map are indicated by numbers, so that their histories can be followed.

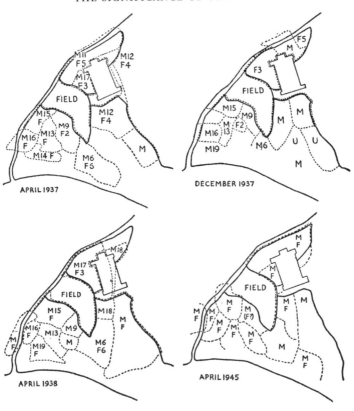

Fig. 6. continued

Where no number is given, the bird had disappeared before the date of the next map.

The census of April 1945 was made with unmarked birds, so may be a little less accurate than the others. However, there were at least as many individuals as shown, and it is unlikely that there were any more, except for the doubtful female indicated by a query. It is interesting that, after so long an interval, the robin population should still be almost exactly the same size.

reasonable to suppose that it has value. That it is a psychological necessity if breeding is to occur was well shown in both aviaries, since in each aviary only that male robin which owned the aviary as its territory, and not the other male, was able to breed. Howard, who, though partially anticipated in his views by earlier writers, must be given full credit for establishing the importance of territory in bird life, considered that territory has survival value for two main reasons, first the acquiring and retention of a mate, and secondly the ensuring of a food supply for the young.

Neither view was entirely new. The earliest reference to food territory in birds was given by Aristotle in his *Historia Animalium*. 'The fact is that a pair of eagles demands an extensive space for its maintenance, and consequently cannot allow other birds to quarter themselves in close neighbourhood.' Again, 'In narrow circumscribed districts where the food would be insufficient for more birds than two, ravens are only found in isolated pairs; when their young are old enough to fly, the parent couple first eject them from the nest and by and by chase them from the neighbourhood.' The statement at the head of this chapter is apparently the only other mention of bird territory by a classical writer.

The next references are given by Olina for the robin, already quoted, and also for the nightingale, which is stated to sing in its territory. Territory was also recognized in the mute swan in the seventeenth century. Next comes Buffon (1771-83), who is the second author to subscribe to the food territory theory: 'The nightingales are also very solitary; ... they select certain tracts, and oppose the encroachments of others on their territory. But this conduct is not occasioned by rivalship, as some have supposed; it is suggested by the solicitude for the maintenance of their young, and regulated by the extent of ground necessary to afford sufficient food. The distances between their nests are much smaller in the rich countries, than in others which reluctantly yield a precarious supply.' On the other hand, Gilbert White did attribute this spacing to rivalship. 'During the amorous season, such a jealousy prevails between the male birds that they can scarcely bear to be together in the same hedge or field … and it is to this spirit of jealousy that I chiefly attribute the equal dispersion of birds in the spring over the face of the country.' About the same period, Oliver Goldsmith introduced the word territory for a bird's defended area. As already noted in Chapter 2, Montagu (1802) was the first to recognize the true function of song in pair-formation. After this, there were a number of casual references to various aspects of territorial behaviour, and then Altum gave a clear statement of the food territory theory in 1868. Eliot Howard independently put forward both Montagu's and Altum's conclusions in his *History of the British Warblers* (1907-14), where they were backed by a large amount of new and original field observations.

The value of territory in pair-formation has not been disputed since Howard enunciated it. Territorial behaviour spaces the males out over the countryside, and males which are unmated sing more loudly than those which have acquired mates. Such behaviour must assist the females to find unmated males. It is true that other birds, such as many finches and ducks, form pairs when in flocks, but this does not refute the value of territory in the pair-formation of those species which possess territories. Since in many species, like the robin, the territory is also defended after pair-formation, Howard suggested that it has the subsidiary function of helping to maintain the pair once formed, since they have a definite headquarters and rivals are driven out. This may well be the case.

As regards the food value of territory there has been much dispute. In the robin and many other song-birds the breeding territory is larger than might have been expected if its sole purpose is to assist the meeting of cock and hen. Howard, E. M. Nicholson, and others have therefore claimed that for each species the size of the territory is fixed and is adapted to the food requirements of the young, and that in this way over-crowding is prevented and an optimum spacing of the birds over the countryside is ensured. I, on the other hand, consider that this claim is mistaken, and that the facts can be explained more simply.

In the first place, the idea of 'optimum spacing' involves a fallacy. Natural selection works through individual sur-vival, and this need not result in what is best for the species as a whole. It is the value of the territory to the individual pair and their brood which is the relevant issue. In recent statements in favour of 'food territory', optimum spacing is no longer claimed. For instance, Mrs Nice considers that ter-ritory prevents 'the interference of other pairs with the or-derly sequence of the nesting cycle', so that 'it is necessary that some species carry on all activities within the territory. In such cases the territory must provide a surplus of food for the pair and young in order to allow for adverse condi-tions.' On the other hand, N. Tinbergen, after showing that in some song-birds part of the food is found outside the ter-ritory, writes: 'It is necessary to recognize that there are many passerines to which the territory is necessary to pro-vide *a certain amount* of food.' Thus while Howard and Nicholson claimed that food-territories have value because they contain the optimum amount of food, Mrs Nice does so because they contain a surplus, and Tinbergen because they contain some only. Such marked differences of opinion make one sceptical of theories about food-territory.

That food is in some way limiting for nestling robins can scarcely be doubted – otherwise the average size of the brood would presumably be increased by natural selection until the food limit was reached. However, it may well be that what matters is the time, and not the space, in which the food is collected, in which case a private feeding territory may have no special advantage. In this connexion pair M17 and F3 in 1937 are of interest, as they had a very small territory (see Fig. 6), and often collected food from their neighbour's ground. While this might suggest that their territory was too small, they nevertheless raised a family of five, the normal number in Devon, so their having to collect food from outside their own territory did not put them at any obvious disadvantage.

Although the robin is efficient in keeping out mating rivals, it is inefficient in preserving its territory from food-trespassers. Poaching robins retreat when attacked, but often return unobtrusively after a few minutes. Indeed, feeding is the only one of its activities for which the robin seems to ignore the boundaries of its own territory. Had the food value of territory been important, one might have expected a different state of affairs. Also the robin, like other territorial species, makes no serious attempt to eject food competitors of other species. The latter are attacked only sporadically, and usually for reasons which have no connexion with territory, as discussed in Chapter 13.

The average size of a robin's territory at Dartington was about 1½ acres. Burkitt obtained the same figure in Ireland. Hence at first sight Howard's claim might seem correct that the size of territory is fixed for the species. Detailed study shows that this is not true. At Dartington the larger robin ter-ritories were about two acres in size; indeed, pairs sometimes held over three acres for a short time, after a neighbour had left or died, but they did not maintain such large areas for long. On the other hand, the smallest breeding territories were only ⅔ acre. There seemed nothing peculiar about the make-up of these small territories, which were in woodland similar to that of the larger territories. Further, the area which formed a complete territory in one year sometimes formed part of a much larger territory in the next year. Fig. 6 shows the marked variations that occurred. Clearly, Howard's claim that size of territory is approximately constant does not hold for the robin, in which the largest stable breeding territories are five times the size of the smallest ones in which successful breeding took place.

Similar variations occur in other song-birds, including several which, like the robin, find most of their food on the breeding ground. The territories of the song-thrush in Finland vary from 3¾ to 14¾ acres, and those of the song-sparrow in Ohio from ½ to 1½ acres. In the Californian wren-tit, the ter-ritories also vary from ½ to 1½ acres, these variations being related not to food supply, but to the length of defended bor-der; territories surrounded on all sides by those of other wren-tits were smaller than those in which part of the bound-ary was formed by the edge of the habitat and so did not have to be defended.

Howard's claim is that, should the breeding density of the robin reach a certain figure, then territorial behaviour prevents a further increase; and the implication is that the ter-ritorial limit is reached not infrequently. Presumably, there-fore, the limit must be close to the average size of a robin territory, which in South Devon is 1½ acres. Suppose that one year, owing to unusually favourable conditions, double the usual number of robins are available. On Howard's view, territorial fights would occur and about half the birds would be ejected, so that the number of breeding pairs would be the same as usual. The Dartington evidence, on the other hand, suggests that most, if not all, the robins would secure breeding territories, these being about half the usual size, since in that case the average size of a breeding territory would still be ¾ acre, which is nearly twice as large as the smallest territory in which Dartington robins have bred successfully. If this is what would happen, and there is at present no reason to think otherwise, then territorial behaviour could not be said to limit the breeding density of the robin under normal conditions.

Of course, as Julian Huxley points out, there may be a limit to the degree to which territories can be compressed. This applies to the robin, since in an aviary thirty feet long and ten or twenty feet wide, only one and not two cock robins could breed. But the aviary was about one-thirtieth of the size of the smallest breeding territory. There is at present no evidence that in the robin the limit of compression of territory is reached under natural conditions, or that if reached, this limited area would contain the minimum quantity of food needed by a pair to raise a brood.

If Howard's view is correct, robins must frequently be dispossessed of their territories. As mentioned in Chapter 4, such dispossessed birds are rarely killed. If they remain alive, then once all the good breeding places have been filled, they have only three available alternatives. First, they could stay around throughout the summer, leading a precarious existence in the territories of others. No such territory-less robins were seen at Dartington. Secondly, they could stay in the same district, but staking out territories on unsuitable ground. There is no evidence that they do this. Thirdly, they might move out altogether, to other areas. However, the ringing returns summarized in Chapter 8 show that, once they have taken up territories, the great majority of robins do not move more than a mile during the rest of their lives. There is, therefore, no evidence that robins are dispossessed of their territories on a large scale, as seems required if Howard's views are correct.

At Dartington, and probably in many other places, the robin population stays roughly constant from year to year. As discussed in the last chapter, it is not known what brings this about, but various factors might be involved. There is no special reason for thinking that it is due to territorial behaviour, particularly as the same phenomenon is found in various non-territorial birds, such as the heron. Territorial behaviour undoubtedly helps to space the robins out locally, since a newcomer meets resistance, and, though a persistent individual can usually establish itself, it is easier to settle on sparsely held than on crowded ground. But the spacing of the available robins is a much more limited claim than Howard's view that territorial behaviour actually fixes the number that can be present. The existing evidence is consistent with the view that the average size of a robin territory remains about the same from year to year because, for reasons unconnected with territory, the population remains at about the same level, and that territorial behaviour merely causes a roughly equal spacing of the available birds.

Howard's view may also be questioned on more general grounds. As he pointed out, various birds of prey, wading birds, and even sea-birds are spaced out fairly evenly over their nesting grounds as a result of territorial behaviour. But many of these species, for instance the harriers, the lapwing, the ringed plover, and the guillemot, also various territorial passerine birds such as the swallow and the crimson-crowned bishop, obtain most or all of their food away from the nesting grounds; on the feeding grounds the birds associate without dispute. This shows that territorial behaviour on the breeding grounds can occur in the absence of feeding territories. Hence an equal spacing of birds over their breeding grounds is not, by itself, an argument in favour of the food-territory concept, and that birds such as the robin and the warblers feed in their breeding territories may be purely incidental.

To conclude, the idea that birds regulate their own population density in accordance with the food supply is picturesque, and has from the time of Aristotle caught the imaginations of part of the zoological public. However, there is no real evidence to support it. Although territorial behaviour roughly spaces out the available pairs, there is no good evidence that it effectively limits the number of pairs that can be present. The size of individual territories is much more variable than might be supposed, and robins do not seem to be ejected as often as might be expected if food territories were essential to their existence. Further, various species hold territories although feeding outside their breeding grounds, while birds like the robin, which feed on their breeding grounds, do not effectively maintain their territories against poaching intruders. I therefore consider that territory is primarily of importance in pair-formation, and perhaps in the maintenance of the pair, and that the fact that some species feed in their territories is incidental and probably without particular significance.

The autumn territory of the robin remains to be discussed. Whereas in song-birds breeding territories are the rule, the possession of an autumn territory is very uncommon. The autumn territory of the robin is an individual territory held singly by a male or, less commonly, by a female. The average size is ¾ acre, but there are marked individual variations, as can be seen from the maps. Clearly the autumn territory has no value in pair-formation or any other breeding behaviour, since the robin does not normally breed in autumn. It might therefore be supposed that it can only be a food territory. But there are considerable objections to this view. Food trespassing is regular, possible food competitors of other species are not driven out, and the size of the autumn territories is even more variable than that of the spring ones. Further, territorial fighting is most vigorous during August and September, when food is abundant, and wanes in November and December, when food is getting scarcer. The fighting among the males increases again in late December and January, but this is clearly correlated with pair-formation and not food, and should there occur a spell of extremely cold weather, it seems that territorial behaviour is partially suppressed, *i.e.* at the one time when food is really scarce the owning robin tolerates other robins feeding in its territory.

While there are no other British birds which sing as regularly in autumn as the robin, quite a lot of singing and some chasing can be observed then. American song-sparrows also take up territories with song and fighting for a short time in the fall, this behaviour being much more pronounced in some years than others, while the territories are abandoned during the winter cold weather, when food is scarce. Again, some starlings take up small territories and form pairs in late autumn, but tend to abandon them in cold weather, while in the blackcock the males, but not the females, visit the display grounds in October and fight much as in spring, except that the performance is less vigorous. In the warblers Howard attributed the autumn singing and chasing to play, but this is an inadequate explanation, since I have even witnessed coition by a chiffchaff in August, and Brewster has observed the same in American swallows and also saw them collecting mud for nests, but dropping it immediately afterwards.

These and many similar instances show that there is frequently a partial revival of the spring breeding behaviour among birds in the autumn, though in most cases this behaviour, which is apparently functionless, dies down before the onset of winter. The view is tentatively suggested here that the autumn territorial behaviour of the robin is similar to that of these other species, save only that it is carried further, that it has no particular value in itself, and is simply due to a partial recurrence of the physiological state of early spring. It is therefore of interest that the robin has twice been found actually breeding in autumn, once in September and once in October, while Burkitt's record of a pair which remained together at this season instead of separating represents an early stage of breeding behaviour, though in this case the pair did not proceed further.

To the above view it must be objected that not only the cock robins but also some hens hold individual autumn territories, and that in the hens this does not repeat a condition of the breeding season. Instead the hens behave like cocks in spring. Recent work suggests an explanation for this. First, experimental injection of male sex hormone into the hen canary causes it to sing like a cock, *i.e.* the female bird may have the potentiality for behaving like a male. Secondly, there is circumstantial evidence for a natural secretion of male sex hormone by the female starling, and to this has been attributed the singing sometimes heard from hen starlings in autumn. The same might apply to the autumn singing of hen robins.

It seems astonishing that such striking behaviour as the song, fighting, and claiming of territory by the robin in autumn should be valueless and due to an apparent accident of physiology, but no other explanation so far offered fits the facts. It may be noted that the robin probably has plenty of time to spare in autumn, as food seems abundant, so this behaviour is unlikely to be harmful. There is also the possibility that this physiological accident has survival value for some quite different reason, and the suggestion may be hazarded, tentatively it must be noted, that the factor inducing territorial behaviour in autumn is that which suppresses migratory behaviour. Evidence discussed in Chapter 8 suggests that the urge to migrate north in spring is associated with the sexual hormones, hence it is conceivable that the latter would also inhibit the urge to migrate south. If those British robins which stay the winter survive better than those which migrate, then an internal factor inducing the birds to stay will have survival value. Possibly this accounts for the temporary revival of breeding behaviour among various British-resident birds in autumn. This must be regarded purely as a suggestion and there are certain difficulties, notably that migrant robins sing in their winter quarters and that autumn sexual behaviour has been observed among purely migratory species such as warblers and swallows. It is interesting that a situation similar to that of the robin is presented by the mocking-bird in California. Here also the males are mainly resident, while of the females most migrate, and those which stay behave like males in spring. In the zoological world one does not normally find convergence in either structure or behaviour unless survival value is associated with it.

Postscript

As noted in the preface, it was not possible in this fourth edition to change my discussion of theoretical points without destroying the whole balance of the book, and I have therefore not changed what I wrote on the significance of territory in the robin twenty years ago. It should be noted, however, that the census figures discussed in Chapter 16 give further support to my view that territorial behaviour does not play an important part in the regulation of robin numbers. I might add that, in my view, no later worker has produced a satisfactory account of the relationship between territorial behaviour and bird numbers, and I am as puzzled by this problem now as I was twenty years ago.

12

Adventures with a Stuffed Robin

As soon as Don Quixote perceived them he exclaimed: 'Fortune is guiding our affairs better than we
could have desired, for look yonder, friend Sancho,
where thirty or more huge giants are revealed, with
whom I intend to do battle and take all their lives…'

    'Take care, sir,' cried Sancho, 'for those we
see yonder are not giants but windmills… And nobody could mistake them but one who had other such
in his head.'

    MIGUEL DE CERVANTES SAAVEDRA: *The Ingenious*
Gentleman Don Quixote of La Mancha (1605)

The preceding chapters have been concerned with the life of
the robin under normal conditions, and the bird's behaviour,
though somewhat stereotyped, seems purposive and well
planned. There has been only a very occasional hint of aberration, as when a parent robin fed the fledgling of another in

preference to her own, or begged her mate to feed her when her mouth was full of food. The present chapter is concerned with certain abnormal conditions, and, as is often the case, a study of the abnormal leads to a much truer understanding of the normal.

That a wild bird responds to a stuffed specimen or model of its own species has been utilized by bird fowlers for some hundreds of years, but by those seriously investigating bird behaviour only since two years before this robin study was commenced. Actually the fact was recorded for the robin by three independent observers during the middle of the last century, but I overlooked these records until long after finishing my own experiments. It was reading the experiments of A. A. Allen and F. M. Chapman on American birds which prompted my purchase, for one shilling, of an exceedingly shop-soiled stuffed specimen of a robin. This has proved as valuable an instrument for investigating bird behaviour as field-glasses or coloured rings; it provided the means of photographing the robin's display, and has also afforded considerable pleasure and amusement.

The first attempts with the mounted robin failed. They were made in winter, the stuffed specimen being placed on a branch in a robin's territory and the bird being chivvied gently until it came near. Nothing happened. Later events showed that it was most unlikely that a robin would notice a stuffed bird under such circumstances. A trespassing robin would, after all, be moving about. The specimen was next set up by a stone to which a wild robin regularly came for food which was put out for it. This bird could not fail to see the specimen, but again nothing happened. As was afterwards discovered, this was an exceptionally mild individual, one of the very few that would never attack a stuffed robin.

Fortunately it was decided to make one more attempt, with the first pair of robins to be found building their nest in March. The stuffed specimen was wired to a branch a foot above the nest, and the hen saw it as soon as she returned with building material. At once she started violent threat posturing, and, when the 'intruding robin' did not retreat, flew at it and struck it with her beak. The cock now arrived and also postured violently, stretching his neck to the fullest and dancing all round the specimen. Such a storm over one shop-soiled specimen had not been anticipated, and, since it had not been attached too securely, it was finally turned upside down by the buffets of the hen, and after some more desultory posturing the pair departed.

The experiment was repeated next day, with a similar result except that the attacks were less violent. But the pair then deserted their nest. The stuffed bird was then erected at the nest of another building pair, with the same result as before except that while the birds postured at it vigorously they did not strike it. This pair also deserted their nest. A third building female did not posture at the specimen, but at once started to buffet it so heartily that I had to rush up in order to save it from destruction. When this bird also deserted her nest, experiments with building robins were given up.

The next attempts were made with incubating robins, which it was hoped, rightly as it turned out, would not desert their nests. But incubating females would not attack the specimen and they quickly went back to their eggs, while the males did not come near the nests.

Full success was achieved only when experiments were started using robins with young which were just beginning to get their feathers. At this stage the parents proved as fierce as at any stage of the breeding cycle, while none deserted their nests. To ensure that the young would come to no harm, warm fine days were selected for the experiments, when it would not matter if the parents were kept away from the young for a few minutes.

Experiments were made at as many robins' nests as possible. The first valuable result was to get protracted views of the threat display, so that it could be described more accurately. A wild intruding robin usually departs so soon as the owner robin postures at it, but the stuffed bird does not, hence the attacking robin postures to the maximum and can be observed at length. For the same reason the specimen provided an opportunity of photographing the robin's threat display, and it could be so arranged that the attacking robin had to pose itself in the open against a suitable background.

As the experiments progressed, it became apparent that there were marked individual differences among robins. A few ignored the stuffed bird altogether, some attacked only feebly and briefly, others for rather longer, and some violently. As the appearance of the specimen was similar in each case, these variations must have been due to differences in individual temperament. The existence of the latter had been suspected before, as some robins seemed so much fiercer than others in their attacks on intruders, but in an attack under natural conditions there is no accurate measure of the amount of previous provocation. The stuffed robin, on the other hand, provides the same degree of provocation for each bird.

Individual robins were found to differ not only in fierceness but in their manner of attack. Some postured only, some started by posturing but changed to direct attack after posturing failed to remove the intruder, while a few omitted posturing altogether and simply struck the specimen. Also, the specimen evoked two main types of posturing from different robins, the one consisting of slow rhythmic swaying of the breast with the feet still, and the other of irregular jerking of the head and breast with rapid movement of the feet. Either type could be violent, moderate, or mild. If the specimen was shown to the same robin on a later occasion, it used the same method of attack or posturing as before, showing that such differences were characteristic for each individual. When watching birds, one tends to assume that all individuals of the same species behave in the same way. Possibly a bird watching human beings would come to the same conclusion. Detailed study shows that, in fact, there are considerable individual differences among robins, though, of course, their behaviour is far less variable than that of humans.

If the stuffed robin is left by a robin's nest with young, the parent birds, however fierce, cease after a time to attack it and come to ignore it. Indeed, they may even use it as a convenient perch from which to approach the nest. If the specimen is then taken away, or if it is taken away long before this stage is reached, and if it is then produced again on a later occasion, the parent robins again start to attack, but less strongly than on the first occasion. If the specimen is produced on a third occasion their response to it is yet weaker, and there soon comes a stage when they cease to attack it at all.

This gradual diminution in the fierceness of a robin's attacks against a particular intruder seems quite general. Thus, in each aviary the owning male's animosity against the other cock gradually declined. The bird mentioned in Chapter 4, which repeatedly attacked its reflection in a window, ceased to do so after a fortnight. If a living robin is placed in a cage in the territory of a wild robin, the latter, however fierce at first, eventually ceases to try to attack it. The same effect was also observed under completely natural conditions. One pair of robins with an exceptionally small territory trespassed for food over the boundary of the next territory much more frequently than is usual, and after a while the attacks of the neighbouring pair became conspicuously mild. The same principle is probably involved both when a newly arrived cock by his persistence carves out a territory among established birds, and also when during pair-formation the newly mated cock comes gradually to tolerate his new hen in the territory.

Under the heading 'Cannibalistic propensity of a redbreast' a most circumstantial account has been published of one robin eating another. An accidental experiment with the stuffed robin showed how this presumably came about. While looking for a good place to photograph a robin, we had placed the specimen on its side on the ground. Here it was discovered by the owner of the territory, who promptly attacked it, despite the bird's unlikely attitude. The same occurred with two other robins, and further experiments showed that a robin attacking a stuffed specimen does not pay much attention to its position and will attack it when it is mounted sideways or upside down. Likewise, the robin claimed to be eating another robin had presumably killed a rival and was continuing its attack, as also happened after a fight described in Chapter 4. The bird's mentality is evidently not properly adjusted to meet the unusual circumstance of a dead robin in the territory. However, since in hard weather the robin sometimes eats dead flesh, it is possible that in the above case attack later changed over to feeding – but there is no reason to think that one robin would kill another in order to eat it.

The stuffed robin finally came to a spectacular end, though this provided the idea for a new line of experiments.

An exceptionally violent hen robin attacked the specimen so strongly that she removed its head. For a moment the bird seemed rather startled, but then continued to attack the headless specimen as violently as before, and it seemed as though it might have demolished it completely if I had not interrupted proceedings. It should be stressed again that a bird's mental equipment is adapted to natural conditions, and that in nature headless robins do not suddenly appear beside the nest.

This observation led to an attempt to see just how much of a stuffed robin was needed for a wild robin to treat it as an intruder and attack it. So the headless specimen was further reduced. First its tail was removed. It was still attacked. Its legs had already been replaced by wires to facilitate fastening it to suitable perches, and as a further step its wings were detached. Many individual robins still attacked it. Finally the whole of the body and back were removed, so that the specimen was reduced simply to the red breast feathers with some white feathers below, these being stitched on to a supporting wire. Half the robins to which this bundle of breast feathers was shown displayed at it with typical threat posturing.

Other experiments followed. Just the head and wings of a dead robin were mounted on a wire, and, though the wings stuck out at an absurd angle, several wild robins attacked this specimen. The head and throat alone, mounted on wire, was also attacked at times. A smaller proportion of robins attacked these bits of specimens than attacked the whole specimen, and their attacks, though unmistakable examples of threat display, were often rather milder than normal, though some were violent.

All the specimens described so far had contained red feathers. Experiments were also made with a complete stuffed juvenile robin, which has a speckled brown breast with no trace of red but is otherwise much like an adult robin. Twelve out of fourteen robins to which this specimen was presented ignored it, but one struck it and one both postured and struck. Similar experiments were tried with a complete stuffed adult robin in which the red breast and white abdomen were painted over with brown ink. This specimen had previously elicited violent attacks from wild robins, but after its red colouring had been changed to dull brown it was no longer attacked by any. It was particularly remarkable to find that wild robins which would not attack this complete robin with a brown instead of red breast, yet exhibited typical threat display to a bundle of red breast feathers lacking head, wings, tail, legs, and even body! These results raise in acute form the problem of how one robin recognizes another, but this is deferred to the next chapter. It may also be mentioned in passing that on a few occasions the stuffed red breast elicited food-begging behaviour from fledgling robins just out of the nest, while the whole specimen with a brown instead of a red breast failed to do this. Unfortunately only very few such experiments could be carried out. The various stuffed specimens used in the experiments are shown in the chapter head on p. 158.

Thirty-three wild cock robins to which the whole stuffed robin was shown treated it as they would treat an intruder robin, by attacking it. Two other cocks behaved differently. The specimen had been placed near a nest containing fledglings, and when the cock noticed it he started as usual with a strong threat posturing, but then suddenly stopped and after a pause mounted the specimen and attempted to mate with it. His own hen now came up and he chased her off violently, as if she were an intruder. In the robin the hen gives the signal for the cock to mount by keeping motionless. The stuffed specimen also keeps motionless, hence one need not be too surprised at a male mounting it, and that he was in a muddle over the identity of his partner is suggested by his subsequent attack on his real mate. A parallel incident has been observed in moorhens by Eliot Howard. A male was driving a strange intruding female from his territory when she suddenly assumed the pre-coitional attitude, at which, although he had previously been attacking hard, the male now mounted and coition followed, after which he continued his attacks and drove the female out of the territory. As Bottom once observed, 'reason and love keep little company together'.

One other cock robin mounted instead of attacking the stuffed specimen, and in this case there was no preliminary threat display. This male actually mounted the specimen on three different days, on one of which his own hen arrived and, perching by him, started to give strong threat display at the specimen. The male, though continuing to mount the specimen, ignored his own hen, thus showing that he was fully aware of her identity, since he would promptly have attacked a strange individual robin which perched beside him. This incident shows the extent to which a bird's behaviour is divided up. Fighting and coition are two separate trains of behaviour, the first elicited at almost all times of the year by a strange robin in the territory, and the second only during a few days in the spring by a motionless robin, which is normally the male's own mate. The stuffed robin, being both strange and motionless, provides the correct external situation for both fighting and coition, a concurrence which does not normally happen in nature, and which, when it is in the mood for coition, the cock robin is not equipped to meet. This cock acted as if his mate were in two places at once, both beside him attacking a trespasser and in front of him inviting him to mount, and this without any apparent mental conflict.

Save that two different trains of behaviour were involved, a somewhat similar incident is described by Eliot Howard. He substituted four blown eggs for the newly hatched young in the nest of a yellow-hammer, and at the same time placed the young in another nest close to the proper nest. At intervals the parent bird brought food to the young in the new nest, but when in the mood to brood them she returned instead to the old nest and brooded the four blown eggs. So here the female acted as if her nest were in two different places; as a container of objects to be fed it was in the new position, but as a container of objects to be brooded it was in the old.

As time went on technique improved, and the stuffed robin was carried about already wired to a six-foot stick which could be quickly stuck in the ground in any suitable place. I also came to learn the sort of places in which a robin would best notice the specimen, and so was able to produce attacks even in winter. It was before breakfast on a cold October morning that the strangest of all the results with a stuffed robin was achieved. The stuffed bird had been erected in the territory of a hen robin previously known to be exceptionally fierce, and for the record time of forty minutes this bird continued to posture, strike, and sing at the specimen. She was still continuing to do so when the sound of the distant breakfast gong caused me to interrupt proceedings by removing the specimen from its perch and walking off. By chance I looked back, to see the hen robin return, hover in the air, and deliver a series of violent pecks at the empty air. I was able to get to the exact place where I had previously stood, so could see that the bird was attacking the identical spot formerly occupied by the specimen. Three more attacks were delivered in rapid succession, but on the last two the bird was about a foot out in position. She then sang hard but returned for a final attack, now three feet out of position, while her violent singing continued for some time longer. As Pliny noted, 'Verily, for mine own part, the more I look into Nature's workes, the sooner am I induced to believe of her, even those things that seem incredible.'

There is no record quite comparable in the whole of bird literature. The nearest are some other experiments with nesting birds. Eliot Howard removed the eggs from a linnet's nest and the bird returned and brooded the vacant nest and worked her feet to turn the eggs which were not there. Similarly treated, a 'reed bunting springs up and down on the bottom of the nest to make the young stretch up for food, and behold, they are not there.' But in these cases at least part of the correct situation, the nest, was still there. In the case of the robin the stuffed bird and its perch had completely gone, and there were no solid objects within five feet. Again, an American observer raised the nest of a sooty tern on a platform about a metre off the ground. When the bird became accustomed to the nest at this height it was raised another metre, and when the tern came back it made several ineffectual attempts to incubate in the air a metre above the ground, *i.e.* in the position which the nest had formerly occupied. In this case the bird had become accustomed to incubate at this height over several days, while the stuffed robin had occupied its position for only forty minutes.

Observation under natural conditions suggests a coherent pattern to the behaviour of the robin; what it does and how it does it follow in a logical and purposive sequence, so that we tend to think that, clothed in a robin's body but retaining a human mind, we should do the same things in much the same way and, therefore, for much the same reasons. We tend to assume that the world that a robin sees is much like the world which we see. Suitable experiments show how false this impression is. A headless, wingless, tailless, legless, and bodiless bundle of red breast feathers appears as a rival to be attacked. The nest can be here if one is in the mood to feed the young, there if in the mood to brood them. One's mate can be both beside one fighting a rival and in front of one inviting to courtship. Even the empty air can contain a rival to be destroyed. The world of a robin is so strange and remote from our experience that into it we can scarcely penetrate, except to see dimly how different it must be from our own. Yet it is a world well adapted to everyday life, and leads under normal conditions to actions which appear rational and which therefore deceive us into assuming that the mind which inspires them is not unlike the human mind.

13

Recognition

Mr Weir was also obliged to turn out a robin, as it
fiercely attacked all the birds in his aviary with any
red in their plumage, but no other kinds; it actually
killed a red-breasted crossbill, and nearly killed a
goldfinch.

CHARLES DARWIN: *The Descent of Man and
Selection in Relation to Sex* (1871)

Into the world of the robin we cannot penetrate. Nevertheless
observation, experiment, and objective analysis can provide
some idea of what is important in its world. This chapter is
concerned with one aspect of this problem, namely with how
a robin recognizes other robins intruding in its territory. That
intruders are recognized is evident enough, since in nature the
owning robin promptly drives from its territory all other
robins which try to enter it, the bird's own mate alone being

excepted. On the other hand, birds of other kinds are not usually attacked, so clearly discrimination is exercised. It is mainly through the robin's mistakes that its recognition of other robins can be investigated.

Occasionally a robin does attack other species. One robin chased a dunnock for three minutes in and out of a woodpile, but more usually such attacks last for only a few seconds. At Dartington, the dunnock is attacked far more often than any other bird, and is chased both in flight and on the ground. Sometimes a dunnock is actually struck, but I have never observed a robin to give its threat display at one. More rarely other species are attacked, but almost always only in flight. Robins were occasionally seen to chase chaffinch, greenfinch, pied and grey wagtails, tree creeper, blue and great tits, and chiffchaff, in fact nearly all small birds of about robin size.

The suggestion that such attacks benefit the robin from the food (or any other) point of view can be ruled out. There are many days when birds of other species are tolerated by the robin, and when they are attacked it is usually very briefly, after which they often return to the robin's territory unmolested. Further, some of the birds concerned, such as the greenfinch, eat quite different foods from the robin.

The other obvious suggestion is that the attacking robin has temporarily mistaken these birds for intruding robins. This seems borne out by the way in which a robin will often break off such an attack half-way through, as if recognizing its error. Moreover, the dunnock, being brown and of similar size and actions, looks much more like a robin than do the other species concerned, and it is also attacked much more often. Observations by others suggest that common and black redstarts are attacked yet more frequently than the dunnock, but as redstarts do not occur round Dartington, I have no personal records. A robin has even attacked a stuffed redstart. In shape and actions, redstarts resemble a robin more closely than does a dunnock, and further, they have red in the plumage. One observer found that a robin attacked black red-starts particularly on seeing the red tail.

But the problem is not simply one of mistaken recognition. A robin's powers of recognition are so good that it can distinguish its own mate from all other individual robins. Therefore its eyesight must be acute, and it is difficult to believe that it could even momentarily mistake for a robin such distinctive birds as a blue tit or a pied wagtail. Nor is it a question of experience. Colour-ringing showed that old robins of three years' experience attacked birds of other species about as frequently as did robins in their first year. Further, an individual robin would attack a dunnock one day, ignore it the next, and attack it again on a later occasion, from which it would seem that ability to recognize a bird of another species varied from day to day.

The experiments with stuffed fragments described in the last chapter raise the problem of recognition more acutely. Since a robin can distinguish its own mate from all other robins, it seems astonishing that it should apparently mistake a motionless, headless, tailless, and wingless bundle of red breast feathers for a living robin.

Much of the confusion and apparent contradiction results from the unwarranted use of the term 'recognition'. As Professor Stout wrote, 'Human language is especially constructed to describe the mental states of human beings, and this means that it is especially constructed so as to mislead when we attempt to describe the working of minds that differ in a great degree from the human.' Recognition describes a complicated mental process of human beings which does not necessarily occur in the mind of the robin. While bird behaviour is no longer described naïvely in human terms, progress is still considerably hampered by the unthinking use of words, such as recognition, which carry hidden and usually false implications as to the nature of a bird's mind.

K. Lorenz, the Austrian worker, concluded that many of a bird's actions do not depend on recognition of particular individuals or on appreciation of the situation as a whole, but on the presence of particular simple and highly characteristic signals, or 'releasers'. Any situation which provides a certain signal may elicit the appropriate response. For instance, in many species the signal for the male to mount the female is provided by the latter remaining motionless. The experiments with the stuffed robin and an observation described in Chapter 6 show that it is unimportant to the male robin whether the bird which remains motionless is his own mate or a strange individual. In nature, of course, the bird's own mate is the only robin which ever provides the correct signal, hence the arrangement works in practice.

This is not to deny that for some of its actions a bird recognizes other birds individually. At first a young herring-gull does not distinguish its parents and, when hungry, pecks at any red spot on a suitable background. (There is a red spot on the parent's beak.) Later, however, it comes to distinguish its parents individually from the other herring gulls in the colony, even at some distance. Again a cock robin soon comes to distinguish his mate individually from strange robins as regards not attacking her; though not as regards mounting her.

The view that many of a bird's actions are elicited by relatively simple signals suggests examining whether there is one highly characteristic signal for the attacking behaviour of the robin. Obviously, however, this is not the case, since a flying pied wagtail or blue tit, a bundle of red breast feathers, and a mounted robin lacking red on the breast have nothing in common as regards appearance. Yet all are at times attacked by a robin. These varied objects agree in only one point: each provides one part of what might be termed the natural 'intruder situation', *i.e.* the situation which under natural conditions elicits attack from the owning robin. Thus the flying blue tit or pied wagtail resembles an intruding robin in that it is a small bird in flight, the red breast is the most conspicuous colour on a robin, and the mounted specimen lacking red on the breast has the shape of a robin. Hence I came to the conclusion that the robin's fighting behaviour ought not to be considered as a unity, and that it was really composed of three separate actions, namely flying in pursuit, threat display, and striking, and that each of these actions had its own signal, flying-in-pursuit being elicited by the sight of a small bird flying away, threat display by the sight of red breast feathers, and striking by an object with the shape of a robin.

Observations bore out such a separation. Except for the dunnock and redstart, birds of other kinds were normally attacked only in flight and then only by the robin flying in pursuit. Stuffed specimens of these other kinds placed near a robin's nest did not elicit any attack. On the other hand, a stuffed or living juvenile robin and a living dunnock, all of which have the shape of a robin, were sometimes struck by the robin, though they did not evoke threat display. Again a stuffed red breast elicited threat display, but was not struck. The clearest example of this division of behaviour occurred with a particularly fierce robin which, when presented with the stuffed juvenile robin lacking red on the breast, struck it repeatedly but did not posture, and when this specimen was removed and the stuffed red breast was substituted, promptly changed over to violent threat posturing and did not strike at all.

Of course, a wild intruding robin provides all three of the above signals, and hence under natural conditions it is difficult to detect the three separate elements in the attacking behaviour of the owning robin. However, close observation showed that when a robin was attacking a living intruder it usually postured if the intruder kept still and flew in pursuit only if it took flight.

Hence, provided the robin's attacking behaviour was separated into three parts, the concept of simple signals afforded a much more satisfactory interpretation of it than any explanation in terms of recognition in the human sense. But further observation showed that the above interpretation was too simple and rigid. While the stuffed juvenile robin, which lacks the red breast, was usually struck by an attacking robin and not postured at, very occasionally this specimen did elicit feeble threat posturing. Again the bundle of red feathers, which normally elicited only threat display, was occasionally struck feebly. And a pied wagtail or blue tit, normally pursued only in flight, was very occasionally struck when perching. So, while the robin's attacking behaviour is undoubtedly divided into three parts, this division is not quite complete, there is some overlapping. Without further experiments it is difficult to say more on this point, except that to interpret a bird's behaviour too mechanically is almost as misleading as to interpret it in human terms.

In general Lorenz's original views on signals or releasers would seem to be over-simplified, though they undoubtedly represent a great advance on previous views. Probably most of a bird's actions are elicited not solely by one characteristic signal but by a large number of features in the external situation, though often one of these may be far more important than the rest, and may even have a much greater effect than all the rest put together, as is the case with the red breast in the attacking behaviour of the robin. In his recent work Lorenz comes to a similar conclusion.

There is a fourth element which must be reckoned as part of the robin's attacking behaviour, namely song, since the owner of a territory sings against intruding robins. Song is evoked from the owning robin primarily by the sound of another robin singing in or close to the territory. But, just as the robin occasionally attacks birds of other species, so, as noted in Chapter 2, the song of a canary or the sound of a saw will occasionally induce a robin to sing, while song can also be induced by the sight of a silent intruding robin. Further, the owning robin often starts to sing without any apparent external provocation. Hence the factors eliciting song are varied.

While the sight of red breast feathers normally elicits threat display from the owning cock robin, there is one particular set of red breast feathers which does not produce this effect, namely the red breast feathers belonging to the bird's own mate. Similarly the hen possesses the shape of a robin but is not struck, and flies away but is not pursued. She has even sung occasionally in her mate's territory without evoking any hostile demonstration. Clearly the cock distinguishes his hen individually, which is a warning against interpreting his behaviour too rigidly, since such individual recognition inhibits the attack normally elicited by any of the four abovementioned signals.

As a matter of fact a robin does sometimes make a mistake, though only momentarily. If its mate perches close to it just after it has made a violent attack on an intruder it sometimes postures briefly at it; but the momentary nature of such attacks is really a further demonstration of how readily it does distinguish its mate. The only occasion when the pair posture violently at each other is during pair-formation (see pp. 62-4), but at this stage the pair are not accustomed to each other. The tension between them normally decreases markedly within a few hours, though mild traces occasionally persist much longer. Thus an observer reported that, up till the time of nest building, one hen gave very mild threat display whenever her mate approached; and in another pair, throughout the summer, the cock's head feathers rose whenever the hen came close.

There is another complication. Red breast feathers normally cause a robin to give its threat display (provided the red breast feathers do not belong to the bird's own mate), but this is true only when the robin is in its own territory. If it is in strange territory, red breast feathers have no such effect, indeed they give rise to quite different behaviour, since they make the robin retreat hastily to its own ground, as described in Chapter 4. To this there is one exception, namely when one robin is trying to claim territory from another. This affords a further caution against interpreting the attacking behaviour too rigidly.

Finally, any interpretation of the fighting behaviour of the robin has to take into account the bird's internal state. Some territorial robins seem generally fiercer than others, while on different occasions the same individual may be fiercer or milder. On two exceptional occasions in spring, involving two different individuals, a robin actually permitted a wild intruding robin to feed unmolested in its territory close to and in full view of it, while another individual's attacks were occasionally so feeble that it failed to make an intruder depart. As already noted, such toleration of trespassers becomes commoner in spells of exceptional cold weather and also in late summer. In such cases as these the owner robin was so mild that its attacking behaviour was not elicited, or was elicited but feebly, by the correct external situation. On the other hand, a robin is sometimes so excited that inappropriate objects such as dunnocks are attacked which are more usually let alone. The existence of marked variations in the fierceness of attack was more clearly shown by the experiments with stuffed specimens, since here each robin was provided with the same external situation.

The factors which influence the internal state, the 'fierceness', of the robin are unknown. Injection of male sex hormone induces a female canary to sing and makes a female night heron fiercer, like a male. Hence male sex hormone may have a similar influence on the robin, though experiments have not been undertaken. It should also be noted that among robins good singing and ferocity do not always go together in the same individual. At times this was the case, but on the other hand some robins were good singers but mild fighters, while some fierce fighters were poor singers. Further, unmated males in general sing much more strongly than mated males, but are much less fierce. Again the hen robin is as fierce in March as she is in autumn, but normally she sings only in autumn and not in spring. The main seasonal changes in song and aggressive behaviour are shown in Fig. 7 (opposite).

It is not justifiable to classify individual robins simply as fierce, mild, and so on, since some which gave the most elaborate threat displays rarely, if ever, struck an intruder, while others which struck hard and often were mild when displaying. Others were fierce or mild in both capacities. From this

Opposite: Fig. 7. *Seasonal changes in song and aggressive behaviour*

SONG

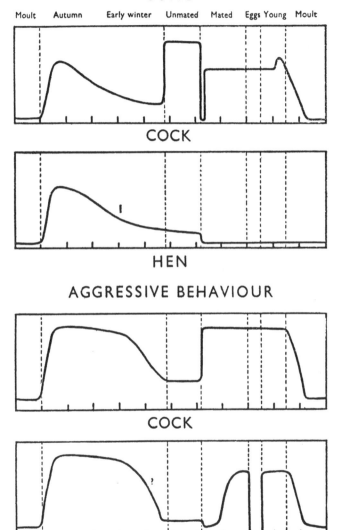

it would seem that the internal state involved in the attacking behaviour of the robin is complicated and does not depend on just one factor, such as the presence of male sex hormone.

Further, external situation and internal state react on each other. Immediately after an attack on an intruder the fierceness of the attacking robin would often seem to be increased, as shown by its occasional attacks at such times on its own mate, who is normally not attacked, and by the bird which attacked the empty air formerly occupied by the stuffed specimen. On the other hand, repetition of the same external situation eliciting attack gradually causes a decline in the bird's fierceness towards a particular object or individual, as discussed in the last chapter.

These observations and experiments show the complexity of the robin's attacking behaviour, which depends on a complicated external situation and a complicated internal state, each of which affects the other, while the way in which the attacks are made also varies. The facts show how misleading it is to interpret bird behaviour either in terms of human behaviour or in terms of a very simple automatic mechanism. But while there is obviously need for a new and better theory, much further work will have to be done before the latter is possible. In particular, the relation between the effective external stimuli and the internal states of the bird needs to be much more fully explored, preferably by experimental methods.

14

Tameness

The redbreast sacred to the household gods,
Wisely regardful of the embroiling sky
In joyless fields and thorny thickets leaves
His shivering mates, and pays to trusted man
His annual visit. Half afraid, he first
Against the window beats; then brisk alights
On the warm hearth; then, hopping o'er the floor,
Eyes all the smiling family askance,
And pecks, and starts, and wonders where he is –
Till, more familiar grown, the table crumbs
Attract his slender feet.

JAMES THOMSON: *Winter* (1726-44)

It is primarily to its tameness in winter that Donne's 'household bird with the red stomacher' owes its place in the affectionate regard of the English people. Indeed, the robin's relation with mankind seems unique. Most birds avoid mankind, usually with too good cause. That this is an artificial and quite unnecessary state of affairs is shown by the well-known fact that the birds of regions uninhabited by man show no such fear. On Bear Island I have picked guillemots off their eggs, and, in the Galapagos, flycatchers have tried to remove the hair of my head as nesting material, a habit accredited to the kite in London in the second century AD.

In winter in Britain, garden birds of many kinds will come to food when it is put out for them, but all of these except the robin tend to retain their habitual shyness and at most come to ignore their benefactor. The robin, on the other hand, is not only remarkably tame but actively seeks out man's habitations at this season, and though this seems entirely inspired by the search for food, there is no other British bird which regularly enters into so domestic a relationship.

Such tameness is chiefly characteristic of the British robin. In Norway, Germany, France, Spain, and Italy, in Corsica and the Canary Islands, most observers describe the robin as a shy woodland species which avoids the haunts of man. In the Alps and Pyrenees, also at Geneva, the household and garden bird is not the robin but the black redstart, and the robin secretes itself in the forests. There are accounts of tame robins in France, Germany, and elsewhere, but they seem to be unusual. This difference in behaviour is brought out when British and Continental robins meet on migration. On the Isle of May, in the Firth of Forth, migrant Continental robins avoid houses and gardens and skulk under banks or among stones, whereas migrant British robins are characteristically tame. On Ushant, off Brest, passage migrants of the British race are again distinguishable by their tameness from the secretive local robins.

The tameness of the British robin is a national compliment. It is not surprising that the bird should be shy in regions where it is hunted and eaten, such as France, Italy, and North Africa, and although the bird is not normally hunted in Scandinavia or Germany, most of the robins from the latter countries migrate to the Mediterranean seaboard, and thus are hunted in winter.

In England it is easy to accustom a robin to feed unconcernedly in the house with people in the same room. Nor is it difficult to get it to take food from the hand. To quote Lord Grey: 'Any male robin can be tamed; such at least is my experience. The bird is first attracted by crumbs of bread thrown on the ground; then a mealworm is thrown to it; then a box – such as one of the small metal boxes in which chemists sell lozenges – is placed open on the ground with mealworms in it. When the bird has become used to this, the next step is to kneel down and place the back of one hand flat upon the ground with the box open on the upturned palm, and the fingers projecting beyond the box. This is the most difficult stage, but robins will risk their lives for mealworms, and the bird will soon face the fingers and stand on them. The final stage, that of getting the bird to come on to the hand when raised above the ground, is easy. The whole process may be a matter of only two or three days in hard weather, when birds are hungry; and when once it has been accomplished the robin does not lose its tameness: confidence has been established and does not diminish when weather becomes mild and food plentiful.'

A spell of hard weather is not the only time when a wild robin can be quickly tamed. Juvenile robins are often extremely tame when they first take up territories in early autumn, and if such a bird is encouraged to enter a house at this season, it need never lose the habit. Correspondents also report that after a tame robin of either sex has acquired a mate in spring, it may come to the tamer accompanied by its mate, and it is then usually easy to tame the latter as well. Should both members of the pair remain in the autumn, they of course occupy separate territories, and conflicts arise when they meet, so the tamer has to use tact as to where he feeds them.

One observer told me of a robin which he had tamed to take food from him. There came a time when he had to shut up his house for half a year. On his return to the garden for the first time in six months he felt a sudden fluttering at his shoulder, and there was his robin. Lord Grey describes a similar incident, so it is clear that a robin can recollect a person after an interval of at least six months. Mrs Nice gives evidence for a song-sparrow remembering an event after eighteen months, while a domestic fowl did so after two years.

The robin's memory is only one of many problems revealed by watching birds which, save for their tameness, were leading natural lives in a wild state. One of the best published descriptions of the robin's threat display was based on observation of a tamed bird, and other incidents observed in a similar way are scattered through this book, relating to courtship-feeding, tension between the pair, food preferences, recognition of young, and other matters. Again, one observer found that his tamed robin took three water-baths daily, even in the coldest weather, a fact which would be hard to establish in a really wild bird. Tamed robins can be watched without disturbance at much closer quarters than is possible with an untamed bird, so that their detailed movements are more easily observed. This valuable method of bird study could profitably be taken up by amateurs more widely than it has been.

Burkitt relates that as he went the round of the robin territories in his Irish garden putting out food, each robin greeted him with song. The meaning of this behaviour is obscure since robin song is normally an indication of hostility. Wild marsh tits fed near Oxford greeted the arrival of their benefactor, but this case is not quite comparable since ordinary call-notes were used. The latter are, of course, regularly used by captive or domestic birds when their human owner appears.

Lord Grey's tame robin would perch on his hand and fill its beak with mealworms when he was standing only a yard from its nest, but the bird would not return to its nest to feed the mealworms to its young until Lord Grey moved some five or six yards away. Another observer tamed both members of a pair of robins to take food from the hand, and when he approached their nest, both members of the pair flew at him and buffeted him. Similarly a correspondent informed me of a pair of robins nesting in the garage of a London suburb which, when the young had hatched, became a considerable nuisance since they attacked all human beings who approached. A neighbour reported the same in a previous year, and another case has been recorded near London. That redoubtable egg-collector the late Rev. F. C. R. Jourdain was once struck in the face by a nesting robin. Again, O. Heinroth found that a captive robin will occasionally attack its keeper.

Much more remarkable is a case related by G. J. Renier. He had tamed a cock robin to take food from him during the winter and it continued to do so during the spring. He eventually saw this cock and its mate visiting their nest, so he approached to look, at which the cock robin perched within a foot of his head and postured violently at him, *i.e.* it treated Renier just as it would treat another robin. No tamer of robins could wish for a greater compliment, even though such treatment was unfriendly. Renier also once observed threat display from his tamed robin when the supply of mealworms temporarily gave out, but further data are needed before this behaviour can be satisfactorily interpreted.

Though so tame in taking food, the robin will not make a pet. Trusting it often becomes, but friendly never. In contrast, a tame raven, parrot, goose, or gull will follow its human owner about with evident friendliness. The reason for the difference is to be found in the natural behaviour of these birds. The birds which make friendly pets are naturally social outside the breeding season; on seeing a fellow member of their own kind they normally associate with it, and their friendliness to mankind is rightly so-called, since it is simply their natural social behaviour directed towards man instead of towards their own kind. The robin, on the other hand, is solitary all the year round, and on meeting a fellow member of its species it normally fights it, hence almost the only part of its behaviour available for transfer to man is hostility. That the latter is occasionally transferred is shown by Renier's tame robin. Izaak Walton's reference to 'the honest Robin who loves mankind both alive and dead' is, unfortunately, as false for the living as it is for the dead.

That Renier's robin could posture at him as though he were a robin raises the problem of how a bird comes to distinguish members of its own species. Formerly it was believed that a bird inherited a knowledge of the distinguishing marks of its own kind. But O. Whitman found that young doves raised from the egg by doves of another species transferred their behaviour to members of the foster-species and later bred with them freely, while they would not breed with members of their own kind. Recognition of their species was not inherited but was acquired at a very early age, after which it could not normally be changed

This discovery has since been confirmed for many other birds and it would therefore have been of considerable interest to know the later history of the robins mentioned in Chapter 7 which were accidentally raised by a blue tit. Did they, for the rest of their lives, vainly pursue blue tits? Lorenz has even found that, when a budgerigar was raised away from all other birds and in company with a suspended celluloid ball, it came to treat the ball as it would another budgerigar, keeping continuously near it, going through the movements of preening it, and later courting it. One courtship movement involves gripping the base of the tail of the other bird, which, the ball having no tail, meant gripping the empty air.

There are also many cases in which captive birds raised by man have transferred their behaviour away from their own kind to man. For instance, hand-reared grey lag geese treated Lorenz as a fellow goose, following him everywhere, and swimming with him in the Danube. Portielje, of the Amsterdam Zoological Gardens, raised a South American bittern which later courted human beings. It bred with a female of its own species only when left alone with it for a considerable time, and even then, if Portielje appeared, would drive the female from the nest and invite its keeper to step into the nest and incubate. Again, under the heading 'A Magpie's delicate attention to its Mistress', the *Literary Gazette* for 12 October, 1850, noted 'A favourite Magpie had been accustomed to receive dainty bits from the mouth of its mistress. The other day it perched as usual on her shoulder, and inserted its beak between her lips; not, as it proved, to receive, for, as one good turn deserves another, the grateful bird dropped an immense green fat caterpillar into the lady's mouth.' In the magpie, as

in the robin, courtship-feeding is regular, and the above behaviour can be interpreted on the assumption that this
courtship-feeding was transferred to man. Perhaps one day a
tamer of robins may be similarly honoured. See also 1 Kings
17:6.

Not all birds transfer their behaviour so easily. Lorenz
states that hand-reared wading birds will not transfer their
behaviour to man. However, an English aviculturist had a
hand-raised tame common snipe which displayed to him, so
perhaps there is individual or specific variation in the matter.
It may also be noted that in those species in which transference seems relatively easy, recognition of their own kind is
not completely undecided at hatching. Thus according to
Lorenz the young grey lag goose treats as a parent the first
moving object which it sees, but a stationary object will not
produce this effect. Hence recognition of at least one characteristic of the parent is inherited. Possibly future work will
show that, as in the case of song, discussed in Chapter 2, each
species has an inherited predisposition towards its own kind,
which may be stronger or weaker in different cases.

For reasons given in the preceding chapter, it is strictly in-
correct to speak of recognition of the species. A bird has various patterns of behaviour, many of which are elicited
primarily by particular signals. When reared by a foster-
species, a bird often comes to respond to signals supplied
by the foster-species. Sometimes, however, the foster-species
cannot supply a sufficient basis for all the correct signals.
Thus Lorenz's tame jackdaw treated human beings as its parents and later courted them, but obviously could not fly with
them, so it flew about with a tame hooded crow, while its
parental behaviour had reference to young of its own kind.

Lorenz at first supposed that once the specific signals had

been determined for the young bird they were fixed irrevocably. This, however, is not so, as shown by Portielje's South American bittern, which was reared by human beings and at first courted them, but eventually bred with a bird of its own kind, though still preferring to court human beings if available. Again, a young grey lag goose taken from its parents, not at hatching but when a few days old, nevertheless came to respond to human beings in preference to birds of its own kind, and a friend had a herring-gull taken from the nest when half-grown which transferred its food-begging and later some of its social behaviour to human beings, but eventually returned to its own kind. A wigeon captured when adult attached itself to a peacock, and a Chinese goose to a dog.

Numerous other records could be cited to show that transference is sometimes possible after the juvenile stage is past, and perhaps the most remarkable case of all was Renier's robin, since this bird was reared by robins and was not tamed until adult, after which it continued to live a normal robin's life in the wild state. Nevertheless on two occasions it gave its threat display at Renier. Clearly the factors underlying the transference of a bird's behaviour from members of its own kind to those of another are far more complicated than first supposed, and many more experiments are needed before this interesting behaviour can be properly understood.

Similar behaviour is also known from mammals. For instance, R. Meinertzhagen writes that a donkey raised with horses would not associate with its own kind, while a young African buffalo raised with cattle always associated with the latter and actually tried to drive wild buffalo away. It is more surprising to find a parallel in insects. It was formerly supposed that the female's knowledge of where to lay its egg was inherited, but recent experiments show that, in some cases at least, this is not so, and that the female seeks to lay in a place with the smell which it experienced as a larva. By subjecting the larva to a different kind of smell, the female insect can be induced to lay in situations widely different from the natural one.

Nor are these the only animals with juvenile fixations. For many years, and often in the face of considerable opposition, psychoanalysts have claimed that much of a man's fundamental emotional behaviour, including his sexual behaviour, is determined in infancy by his relations with his parents. It has even been stated that a man seeks a lover in the likeness of his mother. This, after all, is not so different from what happens in birds. Hence the tameness of birds provides not only pleasure but also a problem of remarkable interest, the solution of which may well assist the better understanding of the behaviour of mankind.

15

A Digression Upon Instinct

They who write on natural history cannot too frequently advert to instinct, that wonderful limited fac-
ulty, which in some instances raises the brute creation,
as it were, above reason, and in others leaves them so
far below it. Philosophers have defined instinct to be
that secret influence by which every species is com-
pelled naturally to pursue, at all times, the same way
or truth, without any teaching or example; whereas
reason, without instruction, would often vary and do
that by many methods which instinct effects by one
alone. Now this maxim must be taken in a qualified
sense; for there are instances in which instinct does
vary and conform to the circumstances of place and
convenience.

GILBERT WHITE: *Letter LVI to Barrington* (1789)

This chapter is concerned with words rather than birds. The

robin's fighting and other behaviour is of the type usually called instinctive. 'Beware instinct. Instinct is a great matter,' Falstaff exclaimed. A friend asked me how swallows found their way to Africa, to which I answered, 'Oh, by instinct,' and he departed satisfied. Yet the most that my statement could mean was that the direction-finding of migratory birds is part of the inherited make-up of the species and is not the result of intelligence. It says nothing about the direction-finding process, which remains a mystery. But man, being always uneasy in the presence of the unknown, has to explain it, so when scientists abolish the gods of the earth, of lightning, and of love, they create instead gravity, electricity, and instinct. Deification is replaced by reification, which is only a little less dangerous and far less picturesque.

The study of behaviour is filled with words which, often introduced as harmless analogies to help the comprehension of new ideas, have gradually acquired entities of their own and finally become explanations. To quote Eliot Howard, 'No sooner do I begin to talk about an instinct or think about it than I find myself visualizing an agent, and my body full of agents urging me to act – fight or fly, love, hate, or wonder. Yet no one knows what it is or where it is; it describes nothing, measures nothing, but treats of an entity or a thing; and whatever we call it – whether force, urge, or impulse – we imply an organism to be forced, urged, or impelled, and so commit ourselves to dual interpretation. There is no force apart from the bird, no bird apart from the force; there are not two organisms but one.'

The word instinct also suffers from popular misusage, for both habits and intuitive notions in man: 'I instinctively disliked him,' 'On seeing the approaching lorry he instinctively reached for the handbrake,' 'He has all the instincts of his class' (referring to a man not a termite). The biological usages have been broad enough, but are hardly sufficient to include these.

Before proceeding further, instinct should be defined, but consultation of biological, psychological, and philosophical works shows that there is no standard definition, and that instinct has been defined by various workers in relation to very different criteria. The definitions, not the definition, of instinct must be considered.

Many definitions of instinct take for their basis a particular theory as to its nature or origin. Herbert Spencer termed it compound reflex action. A wider knowledge of the complexity of instinctive behaviour, both on the perceiving and the responding sides, has caused this definition to be superseded. Another nineteenth-century definition, due to Eimer, was 'inherited habit', arising through the belief that if an animal continually repeated certain behaviour until it became routine, its offspring would inherit such behaviour. There is no evidence that inheritance of this type occurs so this definition also falls. (The more general notion of inherited behaviour is discussed later.) Any definition based on a particular theory is highly unsatisfactory, since it stands or falls by the particular theory, which will inevitably be outmoded sooner or later.

Other definitions of instinct, such as 'action without foreknowledge of the end in view', depend on unobservables. The instinct of processional caterpillars is to follow each other, so that when Fabre enticed some on to the circular edge of a vase, where they formed a closed circle, they followed each other round it for seven days. Birds and insects disturbed in their normal routine often act so inefficiently that it seems that they must be unaware of the end in view. But there is no certain means of finding this out, and a definition which cannot be tested is valueless. The nearer an animal to man, the greater this difficulty becomes: 'When the poor speculatist, with a serious metaphysical pensive face, addressed him, "But really, Sir, when we see a very sensible dog, we don't know what to think of him"; Johnson, rolling with joy at the thought which beamed in his eye, turned quickly round and replied, "True, Sir; and when we see a very foolish fellow, we don't know what to think of *him*."' To modify the definition of instinct to 'action as if without fore-knowledge of the end in view' weakens it considerably, and further the 'as if' is as open to dispute as the previous dogmatic statement.

There have been religious definitions. To man was given intelligence, the knowledge of good and evil, and the power to fall, but to animals a divine instinct which was infallible. (This was before the processional caterpillars found themselves on Fabre's vase.) Cuvier in 1817 wrote of the incubation of birds: 'So much tenderness and trouble lavished without compensation: such a sublime and generous self-devotion in the most urgent dangers, proves that this natural and amiable sentiment is not the result of any mechanical connexion of ideas and sensations, but of a law altogether divine.' But Erasmus Darwin had already written in 1801: 'And this power has been explained to be a *divine something*, a kind of inspiration; while the poor animal, that possesses it, has been thought little better than a *machine*.'

Instinct comes from the Latin 'to incite', and many definitions centre round the idea of an internal entity, force, or urge driving the animal. Certainly the bird often seems to be acting under compulsion, but the idea of a drive, while useful as an analogy, has limited application and is misleading and false when pressed too far. The drive of a machine is put into it from outside by man, but there are insurmountable difficulties in the idea of a drive which both directs a bird and is created by it. The impossibilities inherent in dualistic interpretation are sufficiently exposed by the quotations from Erasmus Darwin and Eliot Howard already given. Further, a drive is another unobservable, which, as Howard points out, introduces mythological interpretation.

A satisfactory definition of instinct must relate not to the way in which it is supposed to work or to have come about but to its observable characteristics. Some of these characteristics were included in the definitions already discussed, but were there mixed with less commendable matter. The characteristics generally considered to be the essential features of instinctive behaviour are: that it is inherited behaviour and not acquired by learning or in any other way during the life of the animal; hence that it is characteristic of the species rather than the individual, and all the individuals of the same species behave in approximately the same way; that it is mainly stereotyped and unmodifiable except in minor detail; further that it is a complex pattern of behaviour involving the whole animal in a series of co-ordinated actions. The last provision is necessary to distinguish instinct from reflex action. Reflexes such as the knee-jerk or eye-blink have some of the characteristics listed above but involve only very limited portions of the body.

Usually another characteristic is added, which provides a further distinction between an instinct and a reflex, namely the dependence of instinct on a particular type of internal state or urge. A bird's behaviour is sometimes described as purposive or goal-seeking, but the latter terms are unsatisfactory, since, as Lorenz has pointed out, two ideas are involved which really have nothing to do with each other. First, there is the goal itself, the survival value of the behaviour, a goal often appreciated by man but very probably not by the bird. A man thinks of a bird as building a nest in order to raise its young, but the bird which is building the nest need have no idea of this goal. Secondly, there is the bird's striving, for which perhaps the best term is 'appetitive behaviour' coined by Craig. Drive, urge, and impulse refer to the same characteristic.

Of the above attributes of instinctive behaviour, that usually considered the most fundamental is that it is inherited. For convenience of discussion a bird's behaviour may be considered under three heads, first the actions involved, secondly the external situation eliciting the actions, and thirdly the internal state of the bird, without which the external situation has no effect. Are these inherited?

Some actions normally termed instinctive are definitely not inherited. Thus as shown in Chapter 2 there are many birds, including the robin, in which the song of the species is for the most part not inherited but acquired when the bird is young. Again, while the general patterns of behaviour tend to be the same for each individual of the species, this is not true for the details of the actions, and the old view that in instinctive behaviour one action automatically follows the next in a rigid chain is definitely false. Modification occurs in relation to both external and internal factors, the latter being illustrated particularly well by those dogs which have lost two of their four legs in accidents and yet have been able to run with considerable speed and accuracy, although the nerves and muscles involved must have been co-ordinated in a quite novel way.

Passing to the external situation, until recently most observers would have said that appreciation of the external situation eliciting an instinctive action was inherited. However, the observations described in Chapter 14 show that in many cases a bird appears not to inherit a knowledge of its own kind, and that, provided it is taken young enough, it can be made to court a human being or a celluloid ball instead. In such cases appreciation of the external situation is highly modifiable. Again, a bird acquires and does not inherit a recognition of some of its natural enemies.

The internal states of a bird predisposing it to instinctive actions can also be changed. As already noted elsewhere, injection of male sex hormone into a female bird may cause it to sing or fight like a male, and injection of prolactin can make a bird broody. Alteration of the external situation can also change a bird's internal state, since if isolated a female bird has been known to adopt male behaviour with song and fighting, while when a turkey cock was tied down to a set of eggs after a few days it became broody like a hen.

These examples are sufficient to show that the statement that instinctive behaviour is inherited must be qualified very considerably, and similar qualifications must be applied to the associated characteristics that it is stereotyped, and typical for the species. Undoubtedly much of a bird's behaviour is predominantly inherited, particularly as regards the main actions involved. Thus the budgerigar which courted the celluloid ball still tried to grip the base of its partner's tail, even though the ball had no tail. Many similar examples are given by Lloyd Morgan. For instance, birds of several species reared in isolation in captivity appeared to have no inherited visual recognition of water, but when they touched it for the first time they went through typical bathing movements, though not always in the water. Again, birds fly, and build typical nests, without instruction or previous experience. In some species the song is inherited. In other birds a large part of the recognition of their species appears to be inherited. In others there is inherited recognition of various kinds of natural enemies. Clearly the extent to which instinctive behaviour is inherited and the extent to which it is acquired during the life, particularly the early life, of the bird vary considerably among different species and for different actions, and much work remains to be done before the extent of the inherited basis of instinctive behaviour will be sufficiently known to incorporate it in a new definition.

The other main characteristic of instinctive behaviour, its dependence on a particular internal state of the bird, was clearly shown in Chapter 13 (pp. 177-80) in the analysis of the robin's attacking behaviour. Lorenz has pointed out that a reflex action is set off solely and invariably in the presence of a particular external situation, and that it can rest indefinitely or continue almost indefinitely without affecting its strength. An instinctive action, on the other hand, weakens with repetition, while the longer its performance is prevented the greater the intensity with which it eventually occurs. If prevented for long enough, the animal will often seek actively for the appropriate external situation, while finally the action may be performed in the absence of this situation. For instance, a captive starling went through the complicated actions of capturing, killing, and swallowing non-existent flies, and a captive humming bird built an imaginary nest.

This appetitive element in a bird's behaviour is extremely difficult to define and describe precisely, to which must be added the further considerable difficulty that it is related to the emotions of the bird. The observer tends to attribute emotions to a courting or fighting bird not only because a man feels emotions under similar circumstances, but because the whole tempo of the bird's actions is speeded up, while it may perform acts such as wing-fluttering or tearing out of grass which are as meaningless as a woman's fluttering her hands or tearing her hair in comparable circumstances. In short, one thinks of the bird as excited, while some would go further and speak of it as angry, jealous, or afraid according to circumstances.

At least in the present state of knowledge, a bird's state of mind is unobservable and unknowable, so it is difficult to attach any real meaning to the statement that, for instance, a robin is jealous. Unfortunately the problem cannot therefore be shelved as irrelevant, since in many cases the bird's emotional state provides the essential clue to the interpretation of its behaviour. Many of the more human anecdotes about birds are rightly doubted, but the following is an authentic account of eider ducks in Iceland: 'At the entrance to the fjord there is a long narrow sand-pit jutting out into the sea, and when the tide is ebbing a very strong current races past the end of this spit into the sea beyond. I watched a party of eiders "shooting" these rapids, and was amazed to see them land on the outer side of the spit, walk across it, and immediately "shoot" the rapids again. They were plainly doing it for enjoyment, and repeated the performance over and over again, sometimes even running across the spit, apparently in great haste to experience the sensation once more.'

Any interpretation of this incident which omitted the emotional state of the eiders would be not only incomplete but misleading. If this is true of a presumably minor emotion, a bird's emotions must play at least as big a part in determining the nature of its militant and amatory behaviour. Hence any description which omits the bird's subjective state of mind is lacking not in an unimportant or irrelevant but in an essential constituent, and since a bird's state of mind seems unobservable and indescribable, the difficulty appears to be insoluble. Certainly it is wiser to omit mention of the bird's state of mind altogether than naïvely to attribute to birds the possession of human emotions which are themselves ill-defined and improperly understood. But it should be stressed that this procedure is adopted not because a bird has no emotions, but because the question of whether it has, and if so what they are like, is, for the present at least, unanswerable.

Perhaps the greatest of all the numerous difficulties in regard to the word instinct is that it was in use considerably before attempts were made to define it. When investigation shows that a new term is needed, it is coined, defined, and used, and if later work shows that the definition is wrong the term is normally abandoned. But this has not happened in the case of instinct. Instinct has come to be associated with the behaviour of certain types of animals, notably birds, insects, and spiders, and it has been defined on the basis of attempts to describe and analyse the behaviour of these animals. Later research has shown these descriptions and definitions to be inadequate, and often they have been wrong. As a consequence the definitions concerned have been abandoned, but the word instinct is still retained, hence confusion as to its meaning is scarcely surprising. The question of whether, for instance, the breeding behaviour of birds is instinctive is, in part at least, a tautology, since the breeding behaviour of birds is taken as an example of instinctive behaviour.

Instinct has often been contrasted with intelligence, but the latter word too has never been defined satisfactorily, hence any attempt at comparison between the two becomes lost in a confusion of indefinite meanings. Rather than compare two abstract and indefinite qualities such as instinct and intelligence, it is much more profitable to compare the analogous behaviour of two types of animal. In a concrete case one can deal with observed facts, and can discuss the degree to which the behaviour of each animal is stereotyped, inherited, modifiable, and so on, and confusion over meanings is avoided.

The term instinct should be abandoned. Too great an importance need not be attached to the difficulty of defining it, since there are many useful words which have not been defined. We could not use thought if we had first to define it. But bird behaviour can be described and analysed without reference to instinct, and not only is the word unnecessary, but it is dangerous because confusing and misleading. Animal psychology is filled with terms which, like instinct, are meaningless, because so many different meanings have been attached to them, or because they refer to unobservables or because, starting as analogies, they have grown into entities. Fortunately, except in so far as they must subconsciously influence his thinking, these terms do not greatly matter to the bird-watcher, who can continue to observe and experiment, and thus in time create a new terminology whose concepts will be clearly defined on the basis of observed facts, and the old confusion will disappear. As Bacon wrote: 'Miris modis homines, more noctuarum, in tenebris suarum acute vident, ad experientiam, tanquam lucem diurnam, nictant et cæcutiunt.' (It is strange how men, like owls, see sharply in the darkness of their own notions, but in the daylight of experience wink and are blinded.)

16

Forest Robins

Since subjects of this kind are inexhaustible.

GILBERT WHITE: *The Natural History
and Antiquities of Selborne* (1789)

My Devon robins lived in small copses and thick hedgerows bordering small fields, in a small orchard, in the scrub that had grown up in a disused quarry and along a stream, and in a small garden; in short, in those intimate habitats, now fast disappearing, which have made the English countryside renowned for its beauty throughout the world. In such places the robin is one of the most typical and abundant of our native birds. But this was not always so, since in primeval times England was covered by a great forest of oaks and other broad-leaved trees. How did the species fare then? This problem was not in my mind when, after moving to Oxford after the war, I studied robins for three years in woods on the nearby Wytham estate, but differences between these woodland robins and those in gardens gradually forced themselves on my attention. Except that the woodland birds were less tame, there were no obvious differences in their behaviour, but they showed important differences in their ecology, particularly in their food, breeding season, clutch-size, nesting success and winter distribution. Since these Wytham robins lived in oak woodland which had happily been left derelict for a long time, conditions were probably like enough to those of the ancient forest for one to be able to see how our robins must formerly have lived.

In summer, robins bred all through the woods, and though they were commonest at the edges and by glades, they occurred deep inside as well. Since, however, the largest wood on the Wytham estate consisted of only 400 acres, I checked this point in the New Forest, much the largest stretch of native forest at that time remaining in England, and here also, robins bred all through it, though commonest at the edges and along rides. The situation was very different in winter, when robins were still not uncommon at the wood edges, but few remained inside, and none did so after hard winters. Whether most of them migrated out of England, or whether they merely moved locally to hedgerows and gardens near at hand is not known, but there is no evidence for any big increase in robins in rural habitats in winter, so probably they migrated. This problem cannot now be solved, as most of the natural forest in the New Forest, and much of the oak woodland in Wytham, have since been felled and replanted. Natural forest is, unfortunately, anathema to the forester, for whom trees are a commercial crop, and woods, as Pope said to Lord Bathurst:

Woods are – not to be too prolix –
Collective bodies of straight sticks.

Since robins presumably occurred chiefly on the edges of
the primeval forest and in openings, it was probably easy for
them, when most of England became cultivated, to spread
from the forest into woods alongside fields, into hedgerows,
and eventually into gardens, where they became tame. That
Chaucer could refer to the 'tame ruddock' shows that the bird
was noted for its tameness already by the fourteenth century,
but apparently it was formerly tame only in winter. Thus in
1544, William Turner described the bird as common round
houses in winter but retiring to woods to breed, and it seems
unlikely that the repetition of this statement by the naturalists
writing in the seventeenth and eighteenth centuries could have
been due merely to uncritical copying. But in the nineteenth century, there are many records of robins nesting in
gardens.

In most parts of the European continent, as mentioned in
Chapter 14, the robin is still described as a shy woodland bird,
and there seem to be three main reasons for this. First, robins
are tame chiefly in winter, and the winters are milder in England than on the Continent at the same latitude, so that more
robins can survive with us in winter. Moreover the soil is regularly dug in England in winter, so that those robins which
are tame enough to follow the man with a spade have a valuable extra source of food. Secondly, where the climate allows
robins to spend the winter on the Continent, chiefly in southern Europe and North Africa, they are regularly killed and
eaten, indeed their migratory fat makes them better eating
than the British birds. Thirdly, cultivated land on the Continent is typically barer than in England, with few hedgerows

or copses, and there are few 'English gardens' in the towns, so that robins have much less cover than in rural England. Various other birds of the European forest, such as the blackbird, song-thrush and dunnock, have also become garden birds, and in these also, like the robin, the change seems to have occurred first in winter, and earlier in England than on the Continent.

There is the further interesting point that in winter, even in England, robins are scarce in woods, though common in gardens and hedgerows. Hence the changes in vegetation due to man must have modified the birds' migratory habits, and presumably a much higher proportion of them spend the winter with us now than could have done in primeval times. The same probably applies to the blackbird and song-thrush, which likewise find much more favourable wintering places in gardens than woods. While the survival of these birds in winter has been improved especially by changes in land-use, the custom of putting out food for birds in winter has probably helped as well, particularly in the robin.

In the rural habitats at Dartington, and also in Oxford gardens, I came to expect the first robins' nests each year near the end of March, and most pairs started laying by the first week of April. It therefore came as a surprise when, despite careful watching, I could find no nests in Wytham woods so early. In 1946, the average date on which the first egg was laid in the first twelve clutches that I found there was 11 April, in 1947, after the cold winter, it was 22 April, and in 1948, much the earliest spring in my twenty years at Oxford, it was 6 April. Why the birds should start laying one or two weeks earlier in gardens than woods is not known, but there is a similar difference in the blackbird. Blackbirds also stop breeding later in gardens than woods, and the same probably applies to robins, as I have seen robins feeding young in late July on several occasions in gardens or beside a country road, but not in woods.

It was noted in Chapter 7 that, in rural England, most robins lay five eggs in a clutch, and though clutches of six occur in May, they are much outnumbered by clutches of five. But there is a difference in woods, not in the size of the first clutches in mid-April, when most robins lay five eggs, but in May, especially in second broods. Of the clutches which I found started during May in 1946 and 1947 in Wytham woods, none contained only five eggs; three contained six eggs (and as one of these also had a cuckoo's egg, and cuckoos often remove one egg of their host, this nest could have started with seven), six contained seven eggs, and one the un-usual clutch of eight. Almost certainly conditions were ab-normal in these two years, since of eight clutches found in Wytham in May in later years by others, six were of six eggs and two of five, while of eight found in May in the Forest of Dean, Gloucestershire, five were of six eggs, two of five and one of seven. Even so, clutches of six were much more fre-quent in these woods than they are in gardens in May. As robins' nests are much more easily found in gardens and road-side banks than in woods, most published records have come from rural areas, and this has concealed the frequency of clutches of six and seven eggs in England in May. The black-bird also has a larger clutch in woods than gardens in May.

On 23 April, 1946, an assistant and I watched a robin's nest with eggs in Wytham from 6 a.m. to 6 p.m. In this time the hen left the eggs 25 times, staying off for spells of between 3 and 10 minutes, her average time off being 6 minutes. The average length of time for which she sat on the eggs was 23 minutes, but she once came off after only 6 minutes, and once stayed on as long as 80 minutes, this being when two people unwittingly stood near the nest, so she was perhaps shy of leaving. Her next longest sit lasted 51 minutes. At another nest, the hen sat for much longer periods, her shortest session being 70 minutes, and her longest 140 minutes, with an average of 96 minutes. During a 10-hour watch, she left the eggs only six times, for an average of 14 minutes, 8 minutes being the shortest and 21 minutes the longest time off the eggs. Hence her spells off were also longer than those of the other hen. It is curious that there should be such big individual variations in this behaviour.

In Wytham in the summers of 1946 and 1947, as many robins' nests as possible were visited each day to weigh the nestlings. The parent birds did not seem to mind. Indeed one hen, instead of flying off, used merely to slip away a few inches down a tunnel in the bracken, returning to the nest as we were leaving. The nestling robins also got used to us. Every bird-ringer has been told not to handle young thrushes or young robins after they are about ten days old, as they squeal, struggle, defecate and, if replaced in the nest, burst from it, and never return, so that they are in extra danger from enemies. But the behaviour of young robins which have been weighed every day from hatching is in remarkable contrast, for when they are ten or even fourteen days old, they do not struggle when handled, or defecate or squeal. Indeed it was usually safe to weigh each of them singly in the open pan of a balance, as it made no attempt to move, but merely watched unconcernedly as it swung up and down. Finally, when such young are put back in the nest, they normally remain quiet and do not attempt to leave. One handling each day is evidently enough to eliminate their alarm when touched, which normally appears so soon as they have grown feathers. This habit of 'exploding' from the nest when touched is, of course, valuable to nestling robins in nature, because the only time when they are likely to be touched is by an enemy about to grab them, and then sudden flight and scattering provide their best chance of escape.

A newly hatched robin weighs between 1.6 and 2.2 grams, about one-fifteenth of an ounce. For the next nine or ten days it grows rapidly, until it weighs some 18 grams, about ten times its weight at hatching and about the same weight as the adult. Between the fifth and eighth days each nestling often puts on two grams a day, and in one brood the average exceeded three grams in one day. From the eleventh day onward there is little further increase in weight; food is brought as often as before, but the birds evidently use it chiefly to grow their feathers.

An unexpected discovery was the marked variation in weight of nestlings of the same age. On the sixth day, for instance, the lightest chick weighed 7.8 grams and the heaviest 14.1 grams, or 80 per cent more, while on the tenth day the lightest weighed 15 grams and the heaviest 22 grams. As this had not been recorded for the young of any song-bird, we at first thought that it might be due in some way to our disturbing the birds, but we later recorded marked variations in young found and weighed for the first time shortly before fledging, so they were clearly natural. Such variations were found both between members of the same brood, showing that the parents did not always share the food equally among them, and also between different broods, showing that some parents brought more food than others, either because they were more skilled or because there was more food near their nest.

Surprisingly, these big variations in weight normally made

no difference to the age at which the nestling robins acquired their feathers or left the nest. The first wing feathers, those of the secondary coverts, normally split from their sheaths when the young are six days old, while the primary feathers do so on the following day, and since this happens almost irrespective of the weight of the birds, it provides a useful way of telling their age in nests found after the young have hatched. Similarly all the young become fully feathered at about the same age, and because of this, all fledglings look the same size; it was only when we weighed them that we found that there were really marked differences. There is an obvious advantage in the young growing at similar speed irrespective of their weight, because this means that when the first chick leaves the nest, the other members of the brood are ready to follow, and their parents can the more easily keep in touch with them.

It is for this reason that the nestling period of the robin is so constant, usually 14, occasionally 13 or 15 days, and a shorter time only if the nest is disturbed. There is, however, one published record of a much longer period, when a cuckoo laid its egg in a robin's nest in so deep a hollow that the young cuckoo could not later eject the nestlings. The parent robins successfully raised both the cuckoo and their own young, but presumably because food was short, their young stayed for as long as three weeks in the nest. This shows that if the young are extremely short of food, their growth may be retarded.

In 1945 and 1946, with an assistant, I made many observations on the frequency with which parent robins fed their nestlings in Wytham. Care was needed to ensure that the parent birds were not worried by the watcher. One pair, which nested close to a busy road, so were used to seeing people, carried on undisturbed when we sat in the open twenty yards away, but those pairs nesting deep in the woods, which rarely saw a man, would not tolerate us at a much greater distance, even when concealed in a hide. The possible error in such work comes not from obviously shy birds, the observations on which can be discounted, but from those which are just a little disturbed by the presence of an observer, as these do not stop feeding their young altogether, but come less frequently; counts thought to have been made in these latter circumstances were rejected, but they are hard to recognize, particularly in those instances where one of the parents behaves normally and the other is somewhat put off.

Both parents bring food to their young from the day that they hatch, and their feeding rate increases as the young get older, partly because the hen spends progressively less time in brooding them, and she stops brooding them altogether after a week, except in heavy rain. We tried to see whether there was any difference in the feeding rates of cock and hen, but the Wytham birds were not colour-ringed and the two sexes could not be reliably distinguished. On several watches the cock, identified by his song, was found to use a different way to the nest from the hen, and sometimes he differed in his feeding routine, but these differences were not consistent, as later in the same watch, or on another day, the birds often changed. During the first two hours of one watch, for instance, the cock normally came in from behind the nest and the hen from one side, but in the course of the third hour the cock (recognized by his song) was found to be using the hen's path, and I did not notice when he changed over. Again, during a four hours' watch at another nest, the parents almost always perched on a small branch close to the nest before feeding their young, but on the next day they did not use it.

Parent robins often bring food at short intervals followed by longer gaps, during which they perhaps find food for themselves. In four hours at one nest, for instance, the parents made 65 visits, about one every 4½ minutes, but there were in this time three intervals of 16, 18 and 19 minutes when they did not come. Feeding may also be interrupted temporarily if a cuckoo, jay, hawk, weasel or other enemy comes near the nest, and also, by a strong wind, especially if accompanied by rain. At two nests watched simultaneously, feeding proceeded actively for four hours, but then ceased completely for 18 minutes during a squall. At another nest watched during a westerly gale with showers, the parents fed the young only three times in one hour. The most rapid rate observed under favourable conditions was 33 feeds in an hour.

Because of these variations, a large series of observations is needed to establish general trends, but the figures suggest that, with young between one and two weeks old, the parents brought food to the nest about 14 times an hour, which meant for the broods in question about three visits per nestling per hour. These observations were made in woods and I do not know if the young are fed so frequently in gardens. This was true of the only brood which I watched in a garden, but at this nest one of the parent birds was tame, and I cheated by providing it with cheese. Once it discovered this source of food, it paid me a number of visits in rapid succession, on each occasion carrying the cheese straight to its young, but then it did not come near me for quite a time. When it eventually returned, it again paid a series of quick visits, on each occasion carrying cheese to its young, and then went off again, and the same thing happened a third time later in the day. This possibly means that robins seek to vary the kinds of food which they bring to their young.

As found for other kinds of birds by earlier workers, the parents of larger broods tend to bring food more often than those with smaller broods, but they do not (presumably because they cannot) increase their feeding rate proportionately to the size of the brood. At the robins' nests watched in Wytham, the average number of visits with food per nestling per hour was about 4 with broods of three or four young, but only about 2½ with broods of five to seven young. This presumably means that each nestling received less food in a larger than in a smaller brood.

In Wytham, nestling robins are fed largely on green caterpillars, which are much the commonest insects at that time of the year in woods. These caterpillars occur chiefly on oaks and other native trees and are far scarcer in gardens. Nestling blackbirds and great tits, which are also fed mainly on caterpillars, are less well nourished in gardens than woods, and the same might well apply to robins in gardens, but I did not study this point. Because of the prevalence of caterpillars in their diet, the naked skin of nestling robins in woods is tinged with what we at first thought was a 'sickly' yellowish-green, unlike the skin of the garden birds, which looks a healthy pink. But the woodland nestlings are not really sickly, it is merely that their skin takes up the green pigment from the caterpillars on which they are fed. D. W. Snow found a similar difference in colour between the nestlings of woodland and garden blackbirds, and I owe this explanation to him.

At intervals, young robins produce a fæcal pellet, white in colour, and neatly contained in a gelatinous envelope. The envelope enables the parents to remove pellets from the nest without breaking them. Conceivably this has hygienic value, but the main advantage is to prevent white fæces from revealing the nest to enemies. Only species with no nest enemies, such as eagles, swifts or kittiwakes, can afford to leave white splashes around their nests. We first noticed fæcal pellets in three-day-old robins, when the pellets weighed 0.1 to 0.3 gram, though one slightly older nestling, itself weighing 7.1 grams, produced a pellet of 0.45 gram, over 6 per cent of its weight. The pellets increase in size as the young grow, and eventually weigh about a gram, the largest being 1¼ gram. One nestling weighing 16.2 gram produced a pellet of 1.2 gram, thus reducing its weight by nearly 7½ per cent. The containing envelope has, of course, no purpose once the fledglings leave the nest, and it then ceases to be made. In one brood, the young no longer had this envelope on their last morning in the nest, but they had stayed for two days longer than usual.

The parents normally remove any pellets in their beaks just after feeding the young, and then carry them some twenty or thirty yards from the nest before dropping them or rubbing them off on a branch. Averaging the results for all our watches at nests with young between one and two weeks old, each nestling produced a pellet once every 77 minutes, but at times the rate was slightly under one an hour. The nestling robins were usually ringed when nine or ten days old. In one family ringed when only six days old, we left each ring rather loose to allow for further growth of the leg. Next day all the rings had gone, presumably pulled off and carried away by the parent birds owing to their resemblance to fæcal pellets.

In nests found in Wytham woods before the young hatched, the number of young which left successfully was 41 out of a possible 63 in 1945, 36 out of 58 in 1947 and 4 out of 7 in two later nests, an average of 63 per cent. This is decidedly lower than the average of 77 per cent for the many nests found by other observers, nearly all in rural surroundings, mentioned in Chapter 9 (p. 128). Nearly all the nestlings lost in Wytham were taken by predators, of which jays were probably the chief. In the blackbird, likewise, the number of nests destroyed by enemies was 86 per cent in Wytham woods as compared with only 50 per cent in Oxford gardens, a bigger difference than in the robin, but a blackbird's nest is more conspicuous. Indeed, considering how hard it is for us to find robins' nests in woods, I am astonished that predators found over a third of those in Wytham during the fortnight that each contained young. As already mentioned, robins' nests are much more conspicuous in gardens and on roadside banks, and the reason that fewer of these are destroyed by natural enemies is presumably because jays hunt chiefly in woods. Hence even if nestling robins are less well fed in gardens, they are probably safer there than in woods.

While, as already noted, robins are as common in the countryside in winter as in summer, they tend to leave the interior of woods in winter. This is presumably because they find it hard to get food inside woods in cold weather, as they feed chiefly on beetles and other small insects in the litter of fallen leaves, and they may not be able to get at them if the litter is frozen hard or covered in snow. In cultivated land, however, the soil is regularly turned by man and robins take advantage of this. We so often see a robin watching someone digging, and this so obviously helps it to get food that we forget to ask how the habit may have originated. But when an animal shows a special relationship with man, it is rarely, if ever, something completely new, and it usually means that the animal has transferred to man a habit evolved in relation to some other animal in the wild. The reason that robins watch gardeners, and that, as the old books recorded, they follow travellers through the forest, was made clear to me in Wytham after a hard frost in January 1946. That morning, several robins living near the edge of the wood came right into the open to feed where scraps had been put out for poultry.

Probably because they were unusually far from cover, they were unusually shy and fled to the bushes on my approach. But the robins inside the wood behaved very differently.

The first that I met dropped several times in quick succession from a bough to the ground, and when I went to the spot, wondering what food it could find, I flushed a cock pheasant which had been scratching through the frozen layer in search of its own food. The pheasant was strong enough to do what the robin could not, and the robin followed in its traces. This robin, unlike those at the edge of the wood, was unusually tame, and when I myself scraped aside a small area of frozen litter, it promptly dropped down to feed there. Later, at three other places in the wood, three different robins came up to me instead of retreating. Now knowing what was expected, I scraped the ground, and each robin promptly fed at my feet. Hence whereas outside the wood the robins treated me as an enemy, inside it I was an agent for breaking the frost layer.

Various of the larger forest animals, such as badgers and in former times wild boar, perhaps help the survival of the robin in hard weather, and one small mammal certainly does so. Two observers, one of whom I later married, sent me records of a robin watching intently above the spot where a mole was tunnelling just under the surface. In both cases the robin followed above the mole, and when a worm was turned up, the robin pounced down and took it before the mole could get it. Hence the robin's association with digging mammals was probably evolved long before the advent of man with a spade.

A more curious example of a wild bird transferring to man behaviour originally evolved in relation to another mammal is that of the South African honey-guide. This thrush-sized bird feeds on the larvæ of wild bees, and even digests wax, but it cannot itself open bees' nests to get at its prey. Instead, after it has found a bees' nest, it seeks out a honey-badger, gives a characteristic call, which the honey-badger recognizes, then leads it to the bees' nest, and after the honey-badger has opened up the nest, feeds on the larvæ. This behaviour is, of course, beneficial to the honey-badger as well, since it is shown the bees' nest. Its relevance here is that the honey-guide has transferred this behaviour from the honey-badger to the African bushmen, given them also its characteristic call and leading them to the bees' nest, which they open; during these proceedings, it may be added, the bushman utters a grunt like a honey-badger.

Each April from 1945 to 1953, I counted the number of singing cock robins in Marley wood and some adjacent plantations in Wytham. After the severe winter of 1944-45, 26 cocks held territories in this area, but in the following year there were twice as many. Then, after the extremely severe winter of 1946-47, the number fell to 35, but after that it rose successively each year to reach 72 in 1951, which was the largest number recorded; in the two following years it fell to 59 and 49 respectively. A longer series of counts of singing cocks was made from 1946 to 1963 in a Surrey oak wood of 40 acres. Here likewise the number fell between 1946 and 1947, from 24½ to 19 (the 'half' meaning a cock whose territory was half outside the wood). The number then rose gradually to 35 in 1953, but fell again in each of the next two years, after rather cold winters, and then fluctuated rather irregularly. Surprisingly, the number rose from 27 to 32 between 1961 and 1962, although this was the coldest winter since 1946-47, but it fell to 21½ in 1963, after the coldest winter for a century. Hence, with one exception, the numbers fell after a hard winter, whereas after a mild winter there was often an increase, though there were several exceptions. The changes in numbers after mild winters did not occur in parallel in the two woods, so were presumably due to local factors.

The average density of cock robins in the 66 acres of Marley wood was three to every ten acres, but it was about twice as high in the small narrow copse which formed part of my study area at Dartington, and also in the Surrey oak wood. It was considerably higher still in the area of scrub and orchard at Dartington, where it reached one to the acre. These differences in density on different types of ground held good each year. They suggest that Marley wood is less suitable for robins than the Dartington copse or the Surrey oak wood, and that an area of scrub and orchard is more suitable than a wood. But, it may be asked, what does 'suitable' mean? Does it mean suitable for raising young? This seems unlikely, because the preferred food for young robins is caterpillars, which are much more numerous in woods than in rough scrub or orchards. Perhaps, therefore, it means suitable for survival in winter, since robins evidently survive less well in woods than rural habitats in winter.

These census figures raise again the question discussed in Chapter 11, of whether, and if so how, territorial behaviour might regulate the numbers of the robin. On one theory, birds regulate the size of their territories in relation to the availability of food for their young. But this can be ruled out for the robin, since territories were larger in oak woods than in orchards or scrub, whereas caterpillars suitable for their young are much more plentiful in oak woods than rural habitats. The blackbird also has much larger territories in oak woods than gardens, though there is much more food for its young in the woods.

On another theory, due to Julian Huxley, territorial boundaries are thought of as acting like a 'partly compressible rubber disc', so that when numbers are high, additional cocks find it increasingly hard to establish themselves. On this view, territorial behaviour sets a nearly, but not quite, constant upper limit to numbers. But this also was not true in Wytham or the Surrey oak wood. Omitting the summers when robins were unusually sparse after a hard winter, the number of territories varied between 41 and 72 in the Wytham area and between 22½ and 33 in the Surrey oak wood; and at least in Wytham, there was no tendency for a more or less fixed upper limit to numbers. The comparison with a 'partly compressible rubber disc' certainly did not hold good in this area.

So far as the available evidence goes, the observed changes in numbers seem most simply explained by supposing that a variable number of robins die each winter, for the most part presumably of starvation, and that in the following spring their territorial behaviour merely spaces out those which are left, without regulating the total number present. But while that is all that the available evidence requires, one is left with the uneasy feeling that there may be more to territorial behaviour than that. If, for instance, all the cock robins in a wood were removed, it is hard to imagine that it would remain empty for long; so the birds perhaps assist in their own dispersion, though how they do so, and in relation to what factors in the environment, is still extremely puzzling.

The observations reviewed in this chapter suggest that woodland as compared with rural robins breed at a rather lower density and survive less well in winter, especially in the interior of woods, while a higher proportion of them migrate (though it is not known how far they go). Further, though the woodland robins lay larger clutches in May than the rural birds, and though many more caterpillars are available for their young, they have a somewhat shorter breeding season and experience heavier losses from predators than the rural birds. As a result, each woodland pair probably raises fewer young in a year than each pair in cultivated land. As Englishmen, we were brought up to believe with William Cowper that 'God made the country and man made the town', and it requires imagination to appreciate that the English countryside is, like the town, an artefact, and that birds live there in unnatural conditions. One also tends to assume that a bird must fare best in its natural habitat, so that it comes as a surprise to discover that robins both raise fewer young, and also survive less well in winter, in their native woods than in gardens or cultivated land. In this respect they are by no means unique, and to cite only one other species, the blackbird is more numerous, raises more young, and survives better in winter, in gardens than in woods. But there are many other English birds to which the reverse applies, and it is to be hoped that the commercial pressures of the present age will not lead to the loss of all our native woods, so that these other species may remain with us.

In conclusion, I would add that, when I came to prepare the fourth edition of this book in 1965, I was surprised to find how few people have published anything new about robins since I ended my own study in 1948. This book was written to stimulate, not inhibit, research, and it would be a shame if it were mistaken for the last word on the subject. So much more awaits discovery, and discovery is so enjoyable, that I hope new readers will act more quickly than their predecessors, for, to quote Gilbert White yet again, 'subjects of this kind are inexhaustible'.

Sweet Amarillis, by a Spring's
Soft and soule melting murmurings,
Slept; and thus sleeping, thither flew
A Robin Red brest; who at view,
Not seeing her at all to stir,
Brought leaves and mosse to cover her:
But while he, perking, there did prie
About the Arch of either eye;
The lid began to let out day;
At which poore Robin flew away:
And seeing her not dead, but all disleav'd;
He chirpt for joy, to see himself disceav'd.

ROBERT HERRICK: *Upon Mrs Eliz: Wheeler,
under the name of Amarillis*

POSTSCRIPT I

In David Lack's footsteps

DAVID HARPER

David Lack decided to 'leave further discoveries to others in the confident trust they will enjoy their researches as much as I have mine'.[1] Obviously, I cannot speak for everyone following in Lack's giant footsteps (sadly, I cannot even mention all studies here), but many of us have had fun.

Here are a just few findings, organized in the order of David Lack's chapters, and cross-referred to the relevant pages of the book. Sources are given in the endnotes on pp. 270-4.

I. MY ROBINS

David Lack did not colour-ring nestling robins (p. 13), which meant that he could not identify juveniles as individuals. Studies marking chicks reveal extra behaviours: e.g. some broods of fledglings are divided between their parents, so that each chick is fed almost exclusively by one adult.[2] Chick and feeder seem to often be of opposite sexes, although frustratingly this depended on finding chicks that had survived long enough to betray their sex behaviourally. Now that robins can be sexed by their DNA,[3] well-funded researchers have some easy pickings!

David Lack could not describe moult in any detail (p. 17), thanks to the paucity of data. We had to wait until 1969 for an estimate of the duration of moult of the primary feathers of about 50 days,[4] subsequently refined to 60 days.[5]

2. SONG

Sonagram analysis reveals that, unknown to David Lack (p. 24), male robins sing more complex songs than females, both when singing voluntarily and when responding to playback. Moreover, both sexes differ in their responses to the songs of males and females.[6]

Emma Brindley[7] found that despite their large song repertoires,[8] robins can discriminate between recordings of neighbours and strangers; the latter represent far more of a threat, and cause a much stronger response.

Robert Thomas and colleagues demonstrated that robins gain less body mass when they sing more; their analyses suggest that this is primarily because they spend less time foraging.[9]

Listen to two robins singing against each other: sometimes they will alternate their songs, as if playing a vocal ping-pong match, but sometimes they overlap them as if arguing over each other. When they overlap their songs, this 'twittering' seems to reflect a willingness to escalate with their opponent.[10]

Artificial lighting has dramatic impacts on this large-eyed songbird. Bart Kempanaers' group found that robins in a quiet, but well-lit, suburb in Vienna began singing at dawn about an hour before those in a nearby unlit forest.[11]

Daytime noise causes robins to sing more at night, independently of nocturnal light levels.[12] There are also other impacts on their song; during the day, individuals in noisier locations sing less complex, shorter songs with a higher minimum pitch than those living in quieter sites.[13] Moreover, when robins hear recordings of song merged with traffic noise, they make similar changes to their own response compared to when the recordings are not merged with noise.[14] Thus, noise pollution affects both the sender of a signal and its receivers.

3. THE RED BREAST

Both sexes display their red breasts during contests (pp. 38-9). Age and sex differences in the breast have been studied in detail.[15] The size of the red breast increases in both sexes between the first and second year; in females it subsequently decreases. The grey fringe around the red breast gets wider with each moult in males, but not females. These trends mean that robins become easier to sex (all things are relative) by plumage as they get older.

4. FIGHTING

Owners usually beat intruders (pp. 47-8): elegant experiments by Joe Tobias support the idea that the high singing rate and low foraging rate of the newcomers, putting them at an energetic disadvantage, is of critical importance.[16] In their first autumn, juveniles only moult some of their plumage (p. 50), excluding the large flight feathers and a variable number of greater coverts.[17] This allows observers to distinguish first-year and adult robins.[18]

Writing during the horrors of the Second World War, David Lack emphasised how pacifist robins are (p. 55). A brutal explanation for most conflicts being resolved without violence is that escalation is incredibly costly. In addition to the obvious risks of injury and death, fighting can also increase the chances of being preyed upon. For example, robins that engaged in combat rather than display were slower to respond to a stuffed raptor.[19]

5. THE FORMATION OF PAIRS

In some populations, robins seem to be more prone to bigamy than David Lack implied (p. 68), with over 5% of males being bigamous.[20] Although the two females often divide the male's territory, and are very aggressive to each other's broods, they can be 'amicable',[21] exceptionally even sharing a nest.[22] Sadly, I have never known the background history of the 'amicable' females that I have met.

6. COURTSHIP

Despite Robert Gillmor's evocative vignette at the chapter head (p. 71), the term 'courtship-feeding' for a male robin feeding a female is rather a misnomer. As David Lack notes most instances do not occur during pair-formation, nor are they followed by copulation (pp. 74-5). Confusingly, I have sometimes seen it occur during the pairing period or immediately before copulation. Somebody needs to invent a better term: 'supplementary feeding'[23] sadly invites confusion with human provision of food.[24]

Marion East found that females with high begging rates laid larger clutches;[25] Joe Tobias and Nathalie Seddon[26] found the reverse, suggesting that the call signals the female's hunger.

Most females call at a high rate early on, suggesting there is an element of 'negotiation' between mates,[27] which may partially explain this difference.

7. NESTS, EGGS, AND YOUNG

More recent information[28] on the topic agrees with this chapter.

David Lack felt that robins were small thrushes that had evolved a chat-like foraging strategy (p. 95). Subsequent biomolecular studies, however, suggest that they are chats not thrushes, and that, moreover, chats are more closely related to Old World flycatchers (family Muscicapidæ) than to thrushes (family Turdidæ).[29] Even before the flood of DNA data, Rolf and Mary Jensen had argued that continued 'separation of chats and flycatchers at family level is unrealistic'.[30]

Alarm calling at the nest (p. 94) has been further studied by Marion East:[31] mammalian predators attract more '*tic*' calls relative to '*seep*' calls than avian predators.

8. MIGRATION

Understanding of dispersal and migration involving Britain and Ireland has improved.[32] David Lack was clearly fascinated by partial migration; studies in Germany reveal that being either resident or 'migrant' are heritable traits.[33]

Juvenile dispersal starts from about three weeks after leaving the nest, although some leave in October after defending a territory near their natal site. Ringing recoveries show that the brown-breasted youngsters can travel several kilometres in the first few days; the fewer longer movements of over

100 km do not occur until the end of the post-juvenile moult (which makes them red-breasted).

Colour-ringing reveals that some adults disperse before moulting and others after – exceptionally as late as November – having defended a territory, and only return the following spring. Most 'migrants' probably do not travel far: only 10% of 2,454 birds present in May and June and recovered between November and February had travelled more than 20 km. Colour-ringing studies suggest that most 'migrants' travel over 1 km. Searching wider areas for marked birds quickly becomes impossible, however: to search up to just 2 km from a breeding site involves traipsing around an area of over 12 ½ square kilometres! Our failure to locate most 'migrants' is particularly worrying since colour-ringing studies reveal that the majority are females: our understanding of the behaviour and ecology of winter females is clearly defective.[34]

Fewer birds ringed in Britain and Ireland are recovered abroad in winter now than in the early twentieth century (combining all data less than 1% of those present in May and June). Similarly, José Luis Tellería felt that the number of robins wintering in Spain had decreased.[35] More study is needed, however, owing to uncertainties about the proportion of birds ringed in breeding areas, and of birds recovered while wintering.[36]

Over much of Europe, there is a large influx of winter visitors from farther north. Strikingly, in an evergreen holm oak *Quercus ilex* forest in SE Spain, breeders seem to vacate the site to be totally replaced by 'migrants' between November and March.[37] Over most of Iberia, however, breeders are largely sedentary despite the arrival of migrants.[38] In a population where most wintering birds were females,[39] males were more likely to gain access to good feeding sites.[40]

Studies of robins have contributed to our understanding of how birds find their way on migration. When robins were first reported to have a magnetic sense,[41] the results were controversial, but further work by Wolfgang and Roswitha Wiltschko[42] confirmed them. A magnetic sense is now suspected to occur in most animals, although in most species it is usually just one of several orientation cues.[43]

Like many birds, robins have two magnetic senses.[44] The first uses the inclination of the Earth's magnetic field, and depends on a flavoprotein called cryptochrome Cry1a, which is located in UV-violet cones in the retina, and absorbs blue light.[45] When it does so, the pigment becomes activated by forming a radical (molecules with an unpaired electron; the missing electron is donated to a receptor called FAD). Activated cryptochrome seems to influence the light sensitivity of the retinal cells, raising the likelihood that robins actually see the magnetic field. This sense depends on the right eye[46] offering clear vision.[47]

The second magnetic sense uses the north-south polarity of the magnetic field (like most human compasses), and relies on crystals of magnetite (an iron oxide) in receptor cells in the upper bill. Reported in robins[48] as recently as 2010, this 'fixed direction compass' has been less studied. Strikingly, the orientation of robins using this sense is influenced by the wavelength of light received by either eye.[49] Quite how the two magnetic senses work together remains unclear.

The intense and prolonged physical activity of migration causes an increased production of reactive oxygen and nitrogen compounds that cause so-called 'oxidative stress'. Oxidative damage can be detected by high levels of protein carbonyls (PCs); a common counter is increased levels of antioxidant enzymes (e.g. glutathione peroxidase, GPx). In autumn both PCs and GPx were significantly higher in robins caught during their nocturnal migration flight than in ones caught during the day while resting.[50] These results showed for the first time that wild birds suffer oxidative stress while migrating, and increase their antioxidant capacity.

9. AGE

James Burkitt's robin seen in Enniskillen 11 years after ringing (p. 119) remains exceptional. The oldest robin in mainland Britain remains the Lancashire bird found dead in 1977 at 8 years, 4 months and 30 days. David Lack may have been overly pessimistic about survival rates (p. 123), especially for youngsters; BTO data suggest annual mortality rates of 58% for adults and 59% for juveniles,[51] giving a typical life-span of nearer two years.

10. FOOD, FEEDING, AND BEING FED UPON

Knowledge of the fruit eaten by robins (p. 132) has expanded both in Britain,[52] and in the western Mediterranean where fruit[53] and acorns[54] are very important in winter. David Lack suggested that robins' large eyes help them to feed earlier at dawn and later at dusk than most songbirds (p. 133). Large-eyed birds certainly sing earlier at dawn.[55]

There are marked differences in how the two sexes of robin forage:[56] females spend longer gleaning for food while hopping on the ground and males do more sallying briefly from perches. The males' behaviour converges on that of females at low temperatures, suggesting a reason for individual territoriality in winter, and for spring pairs separating in cold weather (p. 66).

Domestic cats are indeed a menace to robins (p. 136). For example, nationally nearly a third of ringed robins found dead are cat kills;[57] in Bristol, numbers killed by cats were high compared with robin breeding success.[58]

11. THE SIGNIFICANCE OF TERRITORY

Colour-ringing studies of robins now include the nominate race *rubecula* of continental Europe. Frank Adriaensen and Andre Dhondt found that around Antwerp (Belgium) their territorial behaviour resembled that described by David Lack (pp. 17-20 & 143-50), although fewer females defended winter territories, and these were smaller than those of males.

Mariano Cuadrado found that territorial robins wintering in southern Spain feed in less exposed sites than non-territorial 'floaters'; he suggested that territorial birds are defending dense vegetation, providing safer sites for foraging and roosting.[59] Similarly, using radio-tracking, Ian Johnstone found that wintering territory owners spent over three-quarters of their time in less than 1% of their territory, mainly bushes.[60] The importance of low, dense vegetation to robins is supported by many other studies.[61]

12. ADVENTURES WITH A STUFFED ROBIN

Using stuffed (and now freeze-dried) robins continues to be an important research tool. For instance, David Chantrey and Lance Workman showed that males responded most strongly to stuffed robins if they were 'singing' thanks to playback.[62] The endnotes will reveal other examples.

13. RECOGNITION

Robins often attack other species; most cases may involve defence of food (dunnocks, *Prunella modularis,* can be very annoying, simply by getting in the way, disturbing prey, rather than eating it – illustrated on p. 170), rather than resulting from errors in recognition. Food defence can be seen most easily at bird feeders, but also occurs regularly at rich patches of natural food, such as fruiting European spindle *Euonymus europæus*. In both circumstances, robins have severe problems with aggression from wintering blackcaps *Sylvia atricapilla*.[63]

14. TAMENESS

Taming robins is both spiritually uplifting, and a potentially useful research tool (p. 184); Viscount Grey of Fallodon (depicted on p. 181) provided good instructions, and pointed out some avenues for study.[64] For example, he noted that shortly before breeding, females switch from feeding themselves from proffered food to simply begging for their mate to collect it for them.

15. A DIGRESSION ON INSTINCT

Instinct remains a tricky concept, many authors preferring the term 'innate behaviour',[65] although this is little improvement in the light of David Lack's strictures about reification (pp. 191-3).

16. FOREST ROBINS

David Lack regretted the absence of extensive ancient woodland in Britain (p. 203). Some of the best remaining approximations to primeval European forest can be found in the

Białowieża Forest, straddling the Polish-Bielorussian border. Interestingly, robins there live at much lower density than in Britain, possibly as a result of higher predation and/or a lack of edge habitats.[66] Białowieża robins often nest in tree holes and on the upturned root plates of fallen trees.[67]

Geoffrey Beven's study of a Surrey oakwood (at Bookham Common – pp. 216-7) has been extended, revealing the extent to which woodland interiors are deserted by robins in winter.[68]

David Lack's final paragraph (p. 219), urging further studies of robins, has stood the test of time; there is still a lot to discover. Let us get on with it!

David Harper, Autumn 2015

David Lack at Dartington

The life of 'The Life of the Robin'

PETER LACK

This chapter is about this book itself, and the influence it had on the subsequent life and work of the author, my father David Lack, who died in 1973. Since the original publication of the book in 1943 there have been several other studies of the species. These have added substantially to our knowledge of the bird and have modified some of the conclusions that the original book reached. Hence David Harper, who is one of those responsible for such studies, has contributed a chapter updating both the facts and putting the original into the modern context.

There are still resident robins *Erithacus rubecula* around Dartington Hall, near Totnes, in Devon. They are easily seen and often heard singing as well, and it is possible that those birds are the 70 (or so) times great-grandchildren, probably mainly through the male line, of those which my father studied there in the 1930s.

That is a rather bland and unscientific statement but it actually encompasses some of the more generally interesting scientific things that my father discovered, or proved for the first time, in his study of the robin, which culminated in this book.

There are three main points in the statement:

1) Grandchildren (not grandsons). You cannot tell the sex of a robin in the field from plumage alone, and in the autumn he found that both sexes sing and defend individual territories. Until then it was generally thought that birds, and usually only males, sang to attract a mate, but it was clear from the robin study that the main purpose was actually to keep other birds out of their territory. Indeed the significance of singing and of territory were arguably the major contributions of the whole study to science in general.

2) Mainly through the male line. He found that more females than males dispersed and moved away from their natal area. This was one of the first proven instances of partial migration, with the two sexes migrating to differing extents.

3) 70 times great. Although the robin study was largely about other aspects he did briefly discuss mortality rates, a theme he developed much more later on in his life. He was widely disbelieved when he first proposed that 60% or so of robins died every year, which in turn means that the average length of a generation is not much more than a year.

THE BOOK ITSELF

As with a lot of science, and certainly of biology in the early to middle part of the twentieth century, the initial impetus was nothing to do with any grandiose ideas of furthering science or indeed the potential for writing a book. In September

1933, shortly after graduating from Magdalene College, Cambridge, my father took a job as a biology teacher at Dartington Hall School. (The school closed in 1987.) The robin work started simply as a need for a pastime to keep the boarding children at the school amused in their spare time. It was a very liberal school, especially for the time, and, among other things, the children did not have to attend lessons if they did not want to. Indeed my father relates one incident in his autobiographical essay published in *Ibis* after his death (Lack 1973) where one week a class did not turn up for a double period. So he waited for ten minutes or so, and then went out to look at the robins. The following week he had his own back when he went out to look at robins and the pupils were frustrated. In later weeks that term they were always on time.

The major period of fieldwork on robins was from the autumn of 1934, his second winter at the school, until after the breeding season of 1938. In 1939 he was still on the staff but was actually on sabbatical leave in the Galapagos, a period which resulted in the other book for which he is most widely known in more popular ornithological circles, *Darwin's Finches* (Lack 1947). As he said himself he was anyway getting a bit stale of schoolteaching and was thinking of moving on. Then of course the Second World War intervened in everything.

All the work on the birds, both in the field and the writing up, was done in his spare time. How many schoolteachers today would have time for such even at a private school? The only essential field equipment was colour rings, and this was one of the first studies to use these, although James Burkitt in Enniskillen in Northern Ireland had already done a fairly detailed study using colour rings, also on robins (Burkitt 1924-26). (Burkitt actually only used combinations of black and white rings as he was colour blind. His original maps are now in the archives of the British Trust for Ornithology.)

Most of the actual fieldwork thereafter was simple observation of the birds, describing what they did, although he did build two large aviaries to do some specific experiments on aggressive behaviour.

What is often now forgotten and had several consequences is that the main scientific results of the work were first published in three papers, rather than the book. One described the aggressive behaviour and the influence of the birds' territory (Lack 1939), one was on population changes over four years (Lack 1940a), and the third contained most of the results of the birds in captivity (Lack 1940b). It was only after these appeared that the idea for a book was born. To some extent this involved just rewriting the papers into a more popular style, but he added many other bits to make it a more complete story. The book was finally published by Harry Witherby in 1943, and despite the restrictions imposed by the war on paper supplies etc.

Reading it now it seems to be very descriptive, concentrating predominantly on the behaviour observations he made himself, although there are several places where he relates his observations to biological theories of the time. It is an excellent read however, for what is effectively the first species monograph; by today's standards it is very anecdotal. There are no references quoted explicitly in the text, although there are some extensive notes at the end. The only diagrams and figures are of territory maps and there are very few numbers and certainly no statistics. Rather more surprisingly most of these comments apply to the two main papers as well. In the behaviour paper (Lack 1939) of 50 pages there are only three tables and one of those works out the percentages of the previous table for you. The more ecological paper in *Ibis* has six tables and these would certainly be compressed by today's editors. Similar trends of the increasing use of numbers and less descriptive style of scientific writing are discussed by Snow (1997).

Even at the time it was clearly a very refreshing change from most of the writing available to people interested in birds. Natural history books were not a new invention, but many were fairly turgid, most were mainly anecdotal, and the authors often seemed to get bogged down with detail. What *The Life of the Robin* did was to draw out biological principles clearly and then illustrate them with pertinent examples from personal observation and knowledge.

The book was received very well, at least among the knowledgeable birdwatching fraternity. Bernard Tucker, in a review in *British Birds* (Tucker 1943), said: 'It is everything that a popular exposition by a scientific writer should be, clear, readable, and told in straightforward language, yet losing nothing in accuracy and precision. It is written with great charm and humanity and frequently enlivened with apt and often entertaining quotations or "curious information" from earlier writers, zoological and otherwise.'

I wish the same could be said of much scientific writing today. I wonder how many publishers would tolerate so many of what are effectively just odd quotations in a serious scientific monograph, albeit written in a fairly popular style? Also Tucker's 'accuracy and precision' bit amuses me. As I noted above there are very few numbers and very few statements which would be considered precise in the numerical sense today. This did not change much for a good deal of my father's later work. He more or less avoided formal statistics all his life and, although in later work he often compiled quantitative data in support of his theories, many of these had started out as hunches based on his natural history knowledge. To the annoyance and frustration of his scientific colleagues these hunches too often proved right.

The book proved popular both from a commercial standpoint and among the public. It was fairly quickly followed (1946) by a second edition which incorporated minor changes and a few additions gleaned from correspondents and others. In the archives (held by the Alexander Library of the Edward Grey Institute in Oxford) there is a large and interesting file of correspondence, some solicited and some spontaneous. Several simply relate individual stories, but others are much more detailed and contain a good deal of useful information. They come from all over Britain, and there are also correspondents in Ireland, Holland, Germany (after the war), Finland, France, and Sweden, with several well-known names among them, including Julian Huxley, Maury Meiklejohn, Richard Fitter, Jim Vincent, J. B. S. Haldane and Richard Meinertzhagen.

The requests for further information also provoked a few interesting responses. For example he asked Elsie Leach, then in charge of the bird-ringing scheme based in the Natural History Museum, if he could look through ringing schedules for information on brood sizes. Even then the reply was 'I feel positively exhausted at the mere suggestion of your looking through all the No. 1 cards in search of robins'. He also asked at least some egg collectors for similar information because there is a letter from Edgar Chance which includes: 'Are you sure egg collectors can supply you with the material you require because we select our clutches and do not take them as they come ... Thus one way or another our material will falsify your averages will it not?'

The book itself remained in print and proved popular enough to warrant a paperback edition which was quite a coup for the time. This was published by Pelican Books in 1953 and became the third edition. My father took the opportunity to update and revise several parts of it, although the essence remained the same.

In 1965 a new edition was produced, and it too was reprinted, in the 1970s. As text this was very similar to the Pelican edition except for the addition of an extra chapter on 'forest robins', but it did involve complete resetting in a rather larger format and incorporated redrawing of the figures and a new set of drawings, by Robert Gillmor, in place of the photographs. There was also a later edition in 1970 in the relatively short-lived paperback Fontana New Naturalist series. The present edition uses the text and drawings from the 1965 edition as its base.

All through the 1950s and 1960s there is correspondence from various publishers and potential translators about the possibilities of producing the book in other European languages. English then was not as dominant as it is now among the scientific community, and there was much more perceived need to do such things. French, German, and Swedish were actively considered, the former two more than once. For reasons which are not now clear none of these ever materialised. Certainly one or two samples of translation were not liked either by my father or by Witherbys and the projects folded, but others seemed just to have quietly disappeared. However there are two foreign translations which have been published: Japanese and Hebrew. Why these were considered viable when others were not is unclear.

A completely unrelated use of the book was some time in the 1950s or 1960s, when a short passage was used as a précis and comprehension exercise in a GCE 'O' Level English Language examination paper. If nothing else this suggests it was well written.

ITS LEGACY AND SIGNIFICANCE TO
ROBINS AND SCIENCE

The main legacy to the bird itself was that it re-emphasized the affection in which it was held and it also gave an inkling that robin life was not as law-abiding and peaceful as first impressions suggest. My father realized, as had a few others before and after him, that it is a very aggressive and pugnacious bird; but he certainly did not realize how common really severe fights are, or that 'murder' occurs as often as subsequent workers have found. For more details of this and other subsequent work see David Harper's chapter above.

It is a very tame bird in Britain but it is really only there and in Ireland that it is so. In most of continental Europe it is mainly or entirely a forest bird and keeps well clear of humans. (Perhaps this is why European translations never found suitable publishers?) It is probably this tameness which makes it such a popular bird in Britain and Ireland, but even now most people do not know about its aggressive side.

The real scientific legacy to robins is that nobody then worked on the bird to any extent for nearly forty years. People thought that everything was known, and this was despite a plea in the 1965 edition for people to have another go. When Marion East (eg East 1981) and then especially David Harper (eg Harper 1985) got going though they found out that there was indeed a great deal more to learn, and there still is.

David Harper's is the most comprehensive subsequent study (in the Cambridge Botanic Garden), and it bears many similarities. Both studies used colour rings and were essentially studies of behavioural ecology (although it was not called that in my father's day), with particular reference to the significance of territory. David also had the advantage in that he was doing it full time as a PhD student, and not as an amateur able to work only when classes permitted.

The fundamental significance of a territory for a bird and the significance of song in the maintenance of that territory are probably the most important and lasting scientific legacies of the whole study. Others before my father had noted that robins, and other birds, defended territories, but nobody had really appreciated what they were for. My father was clearly already interested in the subject before he got to Dartington as he co-wrote a major review of the whole topic in 1933 with his own father (Lack and Lack 1933), and he subsequently credited his father with most of the ideas for this.

In addition people had not realized that, in the autumn, both male and female robins defended individual territories, and that, as a consequence, some females (being less dominant) had to move away altogether. This in turn led to ideas about partial migration, an area of study which my father developed much further in the 1950s and was later continued farther still by others. All these ideas were born through detailed observations of individual birds studied over the whole year and for several years, a method in itself which had hardly been used previously.

The Life of the Robin was the first more or less popular account of the life history of a common British bird, and indeed one of the first of any such studies, popular or scientific. Margaret Nice had done a similar job for the song sparrow in the United States (Nice 1937), and James Burkitt had done so for the robin in Ireland (Burkitt 1924-6) but neither of these resulted in a popular account. It was therefore, in effect, the first monograph on a bird in the modern era of science. These have since become commonplace especially with the series published by T. & A. D. Poyser, and the New Naturalist books published by HarperCollins. Even so it is written in a rather more popular style than the average monograph of today, and this is perhaps why it had the influence it had at the time, and continues to have today. Also, because much of the science had already been published, the book was free to be a popular account without having to go too far in the direction of justifying every statement.

SUBSEQUENT WORK

During the Second World War my father had to forget about studies such as these on robins. He was signed up in autumn 1940 to what became the Army Operational Research Group, a group which was primarily involved in work on radar and radio direction finding. Ornithology wasn't completely forgotten, however, as one of his major discoveries during that period was that he (with his friend George Varley) was able to prove that the 'angels' being recorded on radar screens especially in the later stages of the war were actually birds – see Fox and Beasley (2010) for a review of this. My father himself was heavily involved in the development of 'radar ornithology' through the 1950s and this continues today with many studies of migration in particular.

When he came back from the war in October 1945 my father was appointed Director of the Edward Grey Institute in Oxford. Although he was reformulating his ideas on evolution, and how closely related species interact, with his studies on the Darwin's finches of the Galapagos – published in his book *Darwin's Finches* (Lack 1947) – he also wanted to restart fieldwork. Needless to say he started with robins, this time in Wytham Woods just outside Oxford.

His main interests, though, were changing. He was moving away from behaviour studies and on to more purely ecological matters, investigating mortality, the number of eggs and offspring, indeed population regulation in a general sense, although this included how territory might or might not be involved.

Very early in the Wytham study, however, it became very clear that, if you were interested in population dynamics, the robin was actually one of the last species to try to study. To get good data on the breeding and recruitment side of population dynamics you need to find a lot of nests, and unfortunately robin's nests are well known to be among the most difficult to find of any British bird. Very reluctantly therefore he switched the main fieldwork side of his research to other species, in particular the great tit *Parus major,* and in due course the swift *Apus apus*, both of which could be persuaded to nest in boxes which could be inspected and monitored very easily. (In the event much of the great tit fieldwork was done by others, but the work on the nesting of the swift in the tower of the University Museum in Oxford was carried out largely by himself and his field assistant Elizabeth Silva, who became his wife in 1949. Both studies continue to this day.)

It was the work on Darwin's finches which earned him election to the Royal Society in 1951 – he reckoned he was one of the last fellows to be elected on the basis of work done as an amateur – but it is probably the ecological studies of population dynamics and species interactions for which he is now probably most widely known and respected in the scientific community. These also resulted in books: *Natural Regulation of Animal Numbers* (1954), *Population Studies of Birds* (1966), *Ecological Adaptations for Breeding in Birds* (1968) and others. And, although it was *The Life of the Robin* with which he first made his own name, it was *Darwin's Finches* and these others which really put the Edward Grey Institute on the map as one of the foremost places in the world to study birds. However, among the birdwatching fraternity and indeed many of the scientific community the robin book is still his most widely known.

From his own viewpoint the robin study was the last time he studied behaviour to any extent. It was the only time in his life that he ever did any experiments, and he never again did any work on birds in aviaries. However although the ecological books noted above are mainly summaries of the results of fieldwork by others he did continue to work in the field himself. Indeed he was always a 'birdwatcher' and greatly enjoyed just watching birds – for example family holidays almost always involved going to places where new and/or interesting birds could be found.

Sometimes of course this watching led to further interesting science! The swift study noted above led to the book *Swifts in a Tower* (Lack 1956), there was a whole series of papers on migration primarily using radar in the 1950s with much of the fieldwork done during family holidays in East Anglia, and not to mention fieldwork in Jamaica in 1970-1971 which led to his last, posthumously published, book *Island Biology* (Lack 1976).

The extraordinary thing about the Dartington Hall field-work on the robin though was that he did it all in his spare time. Even now, for most of us, the only time we are able to do fieldwork more or less uninterrupted by other things is when doing the research for a PhD. My father did not even have that luxury. He never actually did a PhD.

Changing scientific direction in the late 1940s was not however completely the end of robins. He continued to pub-lish scientifically on the species for several years, some of the papers in conjunction with my mother who continued a spe-cific affection for the species until the end of her life.

Another aspect of robins which he continued was their literary and historical legacy. All the time he had been work-ing at Dartington Hall he had been accumulating as much published material about the species as he could. Some quo-tations from these are scattered throughout the book but he also published a collection of literary references and quotes, both serious and not, in a book *Robin Redbreast* (Lack 1950). At the time it turned out nothing like as popular as either my father or the publishers thought or hoped it would be, but was re-written with substantial additional material in the early 2000s by my brother Andrew (Lack 2007).

And a final part of this non-scientific study of robins was a collection of, mainly Victorian, Christmas cards. Robins ap-peared on many of the earliest Christmas cards sent, and in-deed cards sent for other purposes such as Valentines. The reason is probably to do with the red waistcoat which the postmen of the day wore on duty (and for which they were nicknamed 'robins'). Of course postmen today still have red in the uniform, although the red waistcoat was dropped al-most as soon as the first Christmas cards appeared around 1870. There is also the legend that the robin flew down to comfort Christ on the Cross by singing in His ear, and that the breast was stained by His blood for evermore.

Cards are now very big business especially for Christmas. Birds of all kinds have a major role on many and not just for the bird charities. Casual observation suggests that the robin is, by some margin, still the most popular bird to appear on these, but this habit is largely restricted to British ones (because of their tameness there?). They occur in all kinds of poses and garbs, some appropriate and some totally inappropriate. This has always been so.

Apart from its tameness and the red coat the robin is rather a strange bird to have got this image of peace and understanding at Christmas. As noted above my father received many letters in response to requests for information. In among these are some letters from Richard Meinertzhagen. (My father and he did not exactly see eye to eye on various issues but most of that came later.) A letter dated 1st February 1945 contains comments on various topics, notably weights and the last paragraph notes that the robin has a very large eye which Meinertzhagen suggests is related to its crepuscular habits. The letter then concludes:

'It is certainly the last to bed and earliest up which fits in with his greed and aggression. He has all the characters of Mussolini and I know of no bird less suited to association with the birth of the Messiah.'

Peter Lack, Autumn 2015

This chapter is a modified and updated version of an article which was first published in the magazine *British Wildlife* (Lack 2001), which itself was based on a talk given to a British Ornithologists' Union conference at Dartington Hall in September 1996. Permission from British Wildlife Publishing to reproduce large amounts of this original is gratefully acknowledged.

REFERENCES

Note

In a scientific paper references to the work of others are normally indicated in the text by the author's name, followed by the year of publication. But as such references tend to interrupt the continuity, and have little interest for the general reader, they have usually been omitted in the text of this book. The reader wishing to check the source of any observation in the book not made by myself will find it under the appropriate chapter and page number in the following list. Where more than one reference occurs on the same page of the book, sufficient further guide as to which is meant is provided either by the author's name or by the title of the work, while where any doubt might remain I have added a note on the subject matter concerned. Where the title of a paper is lengthy, I have sometimes given only a short title when it occurs for the second or third time in the list.

Two studies are referred to so often that they are not documented in this way, but are given immediately below:

(i) Burkitt, J. P. (1924-6), 'A study of the robin by means of marked birds', *Brit. Birds*, 17: 294-303; 18: 97-103, 250-7; 19: 120-4; 20: 91-101.
(ii) Nice, M. M. (1937), 'Studies in the life history of the song-sparrow. I', *Trans. Linn. Soc. New York*, 4: 1-247.

Observations from Burkitt's paper always include a reference in the text either to his name or to the place, Enniskillen, where his observations were made. Observations from Mrs Nice's work include mention either of the song-sparrow or of her name.

Observations from my own papers on the robin are also not specified in detail. These papers are:

(i) Lack, D. (1939), 'The behaviour of the robin.
    Part I. The life history, with special reference to aggres
        sive behaviour, sexual behaviour, and territory.
    Part II. A partial analysis of aggressive and recognitional
        behaviour', *Proc. Zool. Soc. Lond.*, 109, A: 169-219.
(ii) Lack, D. (1940), 'The behaviour of the robin. Population
        changes over four years', *Ibis*, 82: 299-324.
(iii) Lack, D. (1940), 'Observations on captive robins', *Brit.
        Birds*, 33: 262-70.
(iv) Lack, D. (1948), 'Notes on the ecology of the robin', *Ibis*,
        90: 252-79.

Detailed References

CHAPTER 1

p. 10 Blake, W. (*c.* 1801-3), 'Auguries of Innocence' from the *Pickering
    MS*. ('A robin red-breast in a cage puts all Heaven in a rage.')
p. 10 Buffon, G., Count de. (1771-83), *Histoire Naturelle des Oiseaux*.
    (This was partly written by Montbeillard. The English edition
    translated by Smellie appeared in 1792. A spurious translation ap-
    peared a few years later and is that usually sold as *Buffon's Birds*.)
p. 10 Nicholson, E. M., and Willson, M. W. (1928), 'The Oxford Trap-
    ping Station', *Brit. Birds*, 21: 290-4. (Describes housetrap, etc.)
p. 11 Drayton, M. (1604), *The Owle*, line 1291.
p. 11 Shakespeare, W. (1623), *Cymbeline*, Act IV, Sc. ii.
p. 12 Thompson, W. (1849), *The Natural History of Ireland*, p. 164
    (robin picked off nest).
p. 12 Ford-Lindsay, H. W. (1911), 'Extraordinary devotion of some
    birds to their nests', *Brit. Birds*, 5: 81-2.
p. 17 Bell, Richard, *My Strange Pets*, pp. 114-15 (quoted by H. S.
    Gladstone (1910), *The Birds of Dumfriesshire*, p. 17). (Captive
    robin killing four others.)
p. 17 Smith, C. (1869), 'Robin and wigeon breeding in confinement',
    Zool., 2, 4: 1865.

CHAPTER 2

p. 23 Holland, Philemon (1601), *The Historie of the World, commonly
    called the Naturall Historie of C. Plinius Secundus*, trans. Phile-
    mon Holland.

p. 23 Cox, Nicholas (1678), *The Gentleman's Recreation.*

p. 24 Owen, J. H. (1914), 'Robin singing at night', *Brit. Birds*, 7: 322.

p. 24 Kerr, H. R. (1937), 'Robin singing at night', *Brit. Birds*, 30: 352.

p. 24 Brand, A. R. (1937), 'Why bird song cannot be described adequately', *Wilson Bull.*, 49: 11-14. (Analysis of sound-tracks of bird song.)

p. 25 Peacock, T. L. (1817), *Nightmare Abbey.*

p. 25 Howard, H. Eliot (1907-14), *A History of the British Warblers.*

p. 26 M(ain), J. (1831), 'Some account of the British songbirds', *Loudon's Mag. Nat. His.*, 4: 412. (Female song in robin.) Also mentioned by E. Jesse in *Gleanings in Natural History*, 2nd series, 1834.

p. 26 Darwin, C. (1871), *The Descent of Man and Selection in relation to Sex.* (Female song in captive robin, skylark, etc., and quotation from Montagu.)

p. 26 Eggebrecht, E. (1937), 'Brutbiologie der Wasseramsel (*Cinclus cinclus aquaticus* [Bechst])', *Journ. f. Ornith.*, 85: 636-76. (Female song in dipper.)

p. 27 Michener, H. and J. R. (1935), 'Mocking-birds, their territories and individualities', *Condor*, 37: 97-140.

p. 27 Miller, A. H. (1931), 'Systematic revision and natural history of the American shrikes (*Lanius*)', *Univ. Calif. Publ. Zool.*, 38: 11-242.

p. 29 Jesse, E. (1834), *Gleanings in Natural History*, 2nd series. (The verse on robin's war-song.)

p. 30 Chaucer, G. (Late fourteenth century), Prologue to the *Canterbury Tales.* (Pardoner's song.)

p. 30 Nice, M. M. (1943), 'Studies in the life-history of the song-sparrow. II', *Trans. Linn. Soc., New York*, 6: 129-32. (Decline in song after pairing.)

p. 30 Montagu, G. (1802), 'Introduction', in *Ornithological Dictionary*, pp. xxviii-xxxiii.

p. 32 Witchell, C. A. (1896), *The Evolution of Bird-Song*, pp. 195, 206-8.

p. 32 Morton, John (1712), *The Natural History of Northamptonshire*, p. 439.

p. 33 Gesner, C. (1555), *Historiæ Animalium III De Avium Naturæ*, pp. 697-9.

p. 33 Syme, P. (1823), *A Treatise on British Song-Birds*, p. 126. (Robin said 'How do ye do'.)

p. 33 Barrington, Hon. Daines (1773), 'Experiments and observations on the sing of birds. Letter to Mathew Maty, Sec. Royal Soc. 1773', published in *Phil. Trans. 63*, and also as appendix V to Thomas Pennant's *British Zoology, Birds*, 4th edition, 1776, vol. II, pp. 561-600 (660-708).

p. 33 Aristotle, *Historia Animalium*, trans. D'A. W. Thompson, 1910.

p. 34 Scott, W. E. D. (1901, 1902, 1904), various titles (the inheritance of song in passerine birds), *Science N. S.*, 14: 522-6; 15: 178-81; 19: 154, 967-9; 20: 282-3.

p. 34 Conradi, E. (1905), 'Song and call-notes of English sparrows when reared by canaries', *Amer. Journ. Psychol.*, 16: 190-8.

p. 34 Huxley, J. (1942), *Evolution: the Modern Synthesis*, pp. 305-7. (Inheritance of song in nightingale and other species.)

p. 34 Morgan, C. Lloyd (1896), *Habit and Instinct*. (Inheritance of song.)

p. 35 Burkitt, J. P. (1918-21), various titles (the relation of song to the nesting of birds), *Irish Nat.*, 27: 140-7; 28: 97-101; 30: 1-10, 113-24. Also (1922) 'Birds' song', *Irish Nat.*, 31: 117-25.

CHAPTER 3

p. 37 Herrick, R. (1648), *Delight in Disorder*.

p. 37 Nine references to posturing of robin:

(i) Roberts, G. (1864), 'Curious habit of the robin', *Zool.*, 22: 9327.

(ii) Ogilvie-Grant, W. R. (1902), *Ibis*, (8) 2: 677-9.

(iii) Aplin, O. V. (1903), *Ibis*, (8) 3: 132-3.

(iv) Kirkman, F. B. (1911), *The British Bird Book*, vol. I, pp. 435-8.

(v) Coward, T. A. (1923), *Birds and their Young*, p. 136.

(vi) Brown, R. H. (1925), 'Field notes from Cumberland, 1924', *Brit. Birds*, 19: 61-2.

(vii) Brown, R. H. (1925), 'The song and courtship of the robin', *Brit. Birds*, 19: 158.

(viii) Dobbrick, L. (1934), 'Rotkehlchenwerbung', *Beitr. z. Fortpflanz. biol. d. Vög.*, 10: 125-7.

(ix) Hendy, E. W. (1941), 'Robin', *Fourteenth Report of Devon Bird-Watching and Preservation Society*, p. 32.

p. 43 Hingston, R. W. G. (1933), *The Meaning of Animal Colour and Adornment*.

p. 43 Six general papers on the origin, content, and meaning of display:

(i) Lorenz, K. (1935), 'Der Kumpan in der Umwelt des Vogels', *Journ. f. Ornith.*, 83: 137-213, 289-413. Also (ii) (1937), 'The companion in the bird's world', *Auk*, 54: 245-73, and (iii) 'Vergleichende Verhaltensforschung', *Zool. Anz. Suppl.* (*Verh. deutsch. zool. Ges.* 41), 12: 69-102.

(iv) Tinbergen, N. (1939), 'On the analysis of social organization among vertebrates, with special reference to birds', *Amer. Midland Nat.*, 21: 210-34.

(v) Lack, D. (1941), 'Some aspects of instinctive behaviour and display in birds', *Ibis*, 407-41.

(vi) Heinroth, O. (1911), 'Beiträge zur Biologie, namentlich Ethologie und Psychologie, der Anatiden', *Verh. V. Int. Orn. Kong.*, 1910: pp. 589-702.

p. 43 Goethe, F. (1937), 'Beobachtungen und Untersuchungen zur Biologie der Silbermöwe (*Larus a. argentatus* Pontopp.) auf der Vogelinsel Memmertsand', *Journ. f. Ornith.*, 85: 1-119. (Grass-plucking by herring-gull.)

p. 45 Hamel, E. D. (1864), 'Light-coloured robin', *Zool.*, 22: 9327.

p. 45 Mouritz, L. B. (1907), 'Ornithological observations in Surrey 1906', *Zool.*, 4, 11: 104. (Freak robin.)

p. 45 Finn, F. (1915), 'Grey-breasted variation in robin', *Zool.*, 4, 19: 436.

CHAPTER 4

p. 51 Lack, D. (1941), 'Notes on territory, fighting, and display in the chaffinch', *Brit. Birds*, 34: 216-19.

p. 51 Tinbergen, N. (1939), 'Field observations of East Greenland birds. II. The behaviour of the snow-bunting (*Plectrophenax nivalis sub-nivalis* [Brehm]) in spring', *Trans. Linn. Soc. New York*, 5: 1-94.

p. 54 Beaumarchais, P. A. Caron de (1784), *La Folle Journée ou le Mariage de Figaro*. (The French quotation is the final line of the play.)

p. 55 Worthington, J. (1884), 'Pugnacity of the robin', *Field*, 64: 586.

p. 55 Morris, F. O. (1853), *A History of British Birds*, vol. III, pp. 114, 124-5. (Other cases of one robin killing another.)

p. 56 Thompson, W. (1849), *The Natural History of Ireland*, p. 165. (Fighting robins picked up by man.)

p. 56 Seebohm, H. (1883), *A History of British Birds*, vol. I, pp. 264-5. (Fighting robins picked up by man.)

p. 56 Waring, S. (1832), *The Minstrelsy of the Woods*, p. 51. (Cat catching fighting robins.)

p. 57 Huxley, J. S. (1914), 'The courtship-habits of the great crested grebe (*Podiceps cristatus*); with an addition to the theory of sexual selection', *Proc. Zool. Soc. Lond.*, 491-562.

p. 58 Jefferies, R. (n.d.), *The Gamekeeper at Home*.

CHAPTER 5

p. 60 Howard, H. E. (1920), *Territory in Bird Life*.

p. 60 Lack, D., and Light, W. (1941), 'Notes on the spring territory of the blackbird', *Brit. Birds*, 35: 47-53. (Pairing in late autumn.)

p. 60 Morley, A. (1941), 'The behaviour of a group of resident British starlings (*Sturnus v. vulgaris* Linn.) from October to March', *The Naturalist*, 788: 55-61.

p. 64 Verwey, J. (1930), 'Die Paarungsbiologie des Fischreihers (*Ardea cinerea* L.)', *Verh. VI. Int. Ornith. Kong.*, 1926, 390-413. Also

(1929) *Zool. Jahr. Abt. f. Allg. Zool.*, v. *Phys. d. Tiere*, 48: 1-120. (Pairing behaviour of common heron.)

p. 64 Allen, A. A. (1914), 'The red-winged blackbird: a study in the ecology of a cat-tail marsh', *Proc. Linn. Soc. New York*, nos. 24, 25: 43-128. (Describes pair-formation.)

p. 65 Roberts, B. (1940), 'The breeding behaviour of penguins', *British Graham Land Expedition 1934-7, Scientific Reports*, vol. 1, no. 3: 195-254. *Brit. Mus. Publ.* (Pair-formation of Gentoo.)

p. 65 Tinbergen, N. (1931), 'Zur Paarungsbiologie der Flussseeschwalbe (*Sterna hirundo hirundo* L.)', *Ardea*, 20: 1-18. (Sex recognition in common tern.)

p. 65 Noble, G. K., and Vogt, W. (1935), 'An experimental study of sex recognition in birds', *Auk* , 52: 278-86. (Experiments on sex recognition with stuffed birds.)

p. 65 Noble, G. K. (1936), 'Courtship and sexual selection of the flicker (*Colaptes auratus luteus*)', *Auk*, 53: 269-82.

p. 65 Tinbergen, N. (1935), 'Field observations of East Greenland birds. I. The behaviour of the red-necked phalarope (*Phalaropus lobatus* [L.]) in spring', *Ardea*, 24: 1-42. (Female recognizes male at close quarters but not at a distance.)

p. 65 Lorenz, K. (1931), 'Beiträge zur Ethologie sozialer Corviden', *Journ. f. Ornith.*, 79: 67-127. (Recognition of each other by jackdaws.)

p. 66 Noble, G. K., Wurm, M., and Schmidt, A. (1938), 'Social behavior in the black-crowned night heron', *Auk*, 55: 7-40.

p. 66 Renier, G. J. (1934), *A Tale of Two Robins*. See also Lack, D., *Brit. Birds*, 32: 23-4. (Cock and hen separating in cold weather.)

p. 66 Colquhoun, M. K. (1940), 'A note on the territorial behaviour of robins during cold weather', *Brit. Birds*, 33: 274-5.

p. 67 Baron, S. (1935), 'Robins changing mates between broods', *Brit. Birds*, 29: 178-9.

p. 67 Hood, T. (Nineteenth century), 'Faithless Nelly Gray: A Pathetic Ballad'. (Ben Battle.)

p. 67 Baldwin, S. P. (1921), 'The marriage relations of the house-wren (*Trogloclytes a. ædon*)', *Auk*, 38: 237-44.

p. 68 Tinbergen, N. (1939), 'The behaviour of the snow-bunting in spring', *Trans. Linn. Soc. New York*, 5: 1-94. (Quotes references to bigamy.)

p. 68 Blanchard, B. D. (1936), 'Continuity of behaviour in the Nuttall white-crowned sparrow', *Condor*, 38: 145-50. (Regular bigamy.)

p. 69 Ryves, Lt.-Col., and Mrs B. H. (1934), 'The breeding habits of the corn-bunting as observed in North Cornwall: with special reference to its polygamous habit', *Brit. Birds*, 28: 2-26.

p. 69 Kluijver, H. N., Ligtvoet, J., van den Ouwelant, C., and Zegwaard, F. (1940), 'De levenswijse van den Winterkoning, *Troglodytes tr.*

troglodytes [L.] *Limosa*', 13: 1-51 (abstract in *Bird Banding*, 1941, 12: 82). (Polygamy in the wren.)

p. 70 Steinfatt, O. (1938), 'Das Brutleben der Sumpfmeise und einige Vergleiche mit dem Brutleben der anderen einheimischen Meisen', *Beitr. z. Fortpflanz. biol. d. Vög.*, 14: 84-9, 137-44. (Marsh-tit pairs for life, and notes on other tits.)

p. 70 Erickson, M. M. (1938), 'Territory, annual cycle, and numbers in a population of wren-tits (*Chamæa fasciata*)', *Univ. Calif. Publ. Zool.*, 42: 247-334. (Pairs for life.)

p. 70 Michener, H. and J. R. (1935), 'Mocking-birds, their territories and individualities', *Condor*, 37: 97-140. (Pair usually separate in autumn.)

p. 70 Lack, D. (1940), 'Pair-formation in birds', *Condor*, 42: 269-86. (Discusses life-pairing.)

CHAPTER 6

p. 72 Howard, H. E. (1929), *An Introduction to the Study of Bird Behaviour*.

p. 73 Lack, D. (1939), 'The display of the blackcock', *Brit. Birds*, 32: 290-303. (Interruption of coition.)

p. 73 Craig, W. (1911), 'Oviposition induced by the male in pigeons', *Journ. Morph.*, 22: 299-305.

p. 73 Craig, W. (1913), 'The stimulation and the inhibition of ovulation in birds and mammals', *Journ. Animal Behav.*, 3: 215-21. (Includes quotation concerning Harvey's parrot.)

p. 73 Matthews, L. H. (1939), 'Visual stimulation and ovulation in pigeons', *Proc. Roy. Soc. Lond.*, B. 126: 557-60.

p. 74 Huxley, J. S. (1914), 'The courtship habits of the great crested grebe (*Podiceps cristatus*), with an addition to the theory of sexual selection', *Proc. Zool. Soc. Lond.*, 491-562.

p. 76 Renier, G. J. (1934), *A Tale of Two Robins*.

p. 77 Aristotle, *Historia Animalium*, trans. D'A. W. Thompson, 1910. (Two female doves forming pair.)

p. 78 Aristotle, *De Generatione Animalium*, trans. A. Platt, 1910.

p. 78 Lack, D. (1940), 'Courtship feeding in birds', *Auk*, 57: 169-78. (This gives references to all the incidents discussed later in this chapter except for those listed below.)

p. 79 Brehm, A. E. (1874), *Bird Life*, trans. H. M. Labouchere and W. Jesse, p. 158. (Injured robin and linnet fed by companion.)

p. 79 Lack, D., and others (1941), 'Courtship feeding in birds', *Auk*, 58: 60. C. Cottam quotes H. Stansbury (1852), *Exploration and Survey of the Valley of the Great Salt Lake of Utah*, p. 193. (Blind white pelican alive at colony.)

p. 79 Rand, A. L. (1942), 'Some notes on bird behaviour', *Bull. Amer. Mus. Nat. Hist.*, 69: 517-24. (Experiments on juvenile shrike.)

p. 80 Shakespeare, W. (1600), *A Midsummer Night's Dream*, Act 1, Sc. i. (Lysander and Hermia.)

CHAPTER 7

p. 82 Winter nests of the robin are cited in the *Zoologist*, 1848, pp. 2019, 2064; 1869, pp. 1566, 1720; 1883, p. 321; 1885, p. 337; 1904, p. 370; also in *Irish Nat.*, 1898, p. 88, and in various County ornithologies.

p. 83 Kirkman, F. B. (1911), *The British Bird Book*, vol. 1: pp. 438-9. (Cock chased from nest by hen.)

p. 83 Frankum, R. G. (1955), 'Nesting material carried by both of a pair of robins', *Brit. Birds*, 48: 235, and Olivier, G. (1959), 'Male robin taking part in nest-construction', *Brit. Birds*, 52: 61.

p. 83 Mitchell, F. S., 2nd ed. H. Saunders (1892), *The Birds of Lancashire*, p. 15. (Nest in dead cat.)

p. 83 Alexander, H. G. (1929), 'Robin nesting on a bed in an occupied room', *Brit. Birds*, 23: 37-8.

p. 83 Medlicott, W. S. (1910), 'Rapid nest-building by a robin', *Brit. Birds*, 4-17.

p. 83 Morris, F. O. (1853), *A History of British Birds*, vol. III, pp. 129-30. (Robin's nest in wagon.)

p. 83 *Evening Standard* (1905), 'Robin's nest on wagon-axle', *Zool.*, 4, 9: 198.

p. 84 Hyde-Parker, T. (1942), 'Strange nests and nesting places', *Bird Notes and News*, 20: 64. (In car and aircraft.)

p. 84 Owen, T. (1912), 'A season with the birds of Anglesey and North Carnarvonshire', *Zool.*, 4, 16: 310. (Robin's nest in woodpecker hole.)

p. 84 Lilford, Lord (1895), *Birds of Northamptonshire*, vol. 1, p. 104. (Robin ousting nightingale and redstart.)

p. 84 Passler, P. W. (1856), 'Die Brutvögel Anhalts', *Journ. f. Ornith.*, 4: 49. (Robin and willow warbler in same nest.)

p. 84 Knight, C. W. R. (1925), *Aristocrats of the Air*, p. 136. (Robin and pied wagtail using same nest.)

p. 84 Balsac, H. de (1926), 'Un cas de parasitisme chez le Rougequeue *Ph. ph. phoenicurus* (L.)', *Rev. Franc. d'Ornith.*, 10: 137-8. (Robin and redstart together.)

p. 84 Steele-Elliott, J. (n.d.), *The Vertebrate Fauna of Bedfordshire*. (Robin nesting with great tit.)

p. 84 Hemming, D. J. (1936), quoted by Boyd, A. W., 'Report on the swallow enquiry, 1935', *Brit. Birds*, 30: 107.

p. 84 Appleby, L. (1860), 'Robins and titmice reared in one nest', *Zool.*, 18: 7171.

p. 84 Other cases of robin and another species using same nest given in private letters from E. P. Chance, G. Charteris, J. H. Owen, B. H. Ryves, N. Tracy.

p. 85 Owen, J. H. (1911), 'A pair of robins building many nests', *Brit. Birds*, 5: 132 and 7: 346. (In stack of pipes.)

p. 85 Battersby, E., private letter on nests in sewage filter.

p. 85 Forrest, H. E. (1911), 'A pair of robins building many nests', *Brit. Birds*, 5: 166.

p. 85 Turner, W. (1544), *Avium præcipuarum, quarum apud Plinium et Aristotelem mentio est, brevis et succinta historia.* See also Evans, A. H. (1903), *Turner on Birds*, p. 157.

p. 86 Ellison, A. (1909), 'Bird-life in a spring snowstorm', *Brit. Birds*, 2: 303.

p. 86 Palmer-Smith, M. (1957), 'Robins rearing two broods in an unusual "screened" nest', *Brit. Birds*, 50: 492.

p. 87 Kirkman, F. B. (1911), *The British Bird Book*, vol. I, 6: 439.

p. 87 Hudson, F. L. (1957), 'Robin's nest of unusual construction', *Brit. Birds*, 50: 124.

p. 87 Bentham, C. H. (1904), 'Robin nesting in tree and hedge', *Zool.*, 4, 8: 263.

p. 87 Ellison, A. (1904), 'Robin nesting in a tree', *Zool.*, 4, 8: 190.

p. 87 Paterson, J. (1908), 'Report on Scottish ornithology for 1907', *Ann. Scot. Nat. Hist.*, 1908: 134. (Domed nest of robin.)

p. 87 Brown, R. H. (1924), 'Field notes from Cumberland', *Brit. Birds*, 17: 225. (Eggs covered with oak leaves.)

p. 87 Harvey, G. H. (1924), 'Field notes from West Cornwall', *Brit. Birds*, 18: 166. (Eggs buried in nest-lining.)

p. 87 Owen, J. H. (1927), 'Large clutches of robin's eggs', *Brit. Birds*, 21: 64-5.

p. 87 Shove, H. W. (1944), 'Extraordinary egg production by a robin', *Brit. Birds*, 38: 117.

p. 87 Philips, C. L. (1887), 'Egg-laying extraordinary in *Colaptes auratus*', *Auk*, 4: 346. (Flicker.)

p. 87 Davis, D. E. (1942), 'Number of eggs laid by herring gulls', *Auk*, 59: 549-54.

p. 87 Aflalo, F. G. (1898), *A Sketch of the Natural History (Vertebrates) of the British Islands*, pp. 138-40. (Removal of eggs causes robin to lay more.)

p. 88 Ryves, B. H. (1928), 'Robin's eggs hatched after desertion', *Brit. Birds*, 22: 63. (Reared eight young.)

p. 88 Emlen, J. T. (1941), 'An experimental analysis of the breeding cycle of the tricolored redwing', *Condor*, 43: 209-19. (Experiments on effect of removing or adding eggs, with other references to same subject.)

p. 91 Hale, J. R. (1930), *Bull. Brit. Oological Ass.*, 2: 30. (Eggs of same type in Kent orchard.)

p. 91 Owen, J. H. (1927), 'Heredity in abnormal egg-colouration', *Brit. Birds*, 20: 273.

p. 91 Punnett, R. C., and Bailey, P. G. (1920), 'Genetic studies in poultry. II. Inheritance of egg-colour and broodiness', *J. Genetics*, 10: 277-99.

p. 93 Jourdain, F. C. R. (1925), 'A study on parasitism in the cuckoo', *Proc. Zool. Soc. Lond.*, 639-67. (Quotes Capek for Moravian cuckoos.)

p. 93 Thompson, W. (1849), *The Natural History of Ireland*, p. 163. (Incubated five weeks; similar case in letter from J. H. Owen.)

p. 94 Cott, H. B. (1940), *Adaptive Coloration in Animals*, p. 128.

p. 97 Butler, A. G. (1874), 'Robin feeding young thrushes', *Zool.*, 2, 9: 4033-4.

p. 97 No author (1833), 'Is the robin known to possess sympathy for other birds as ascribed to it in this paragraph?', *Loudon's Mag. Nat. Hist.*, 6: 69-70.

p. 97 Paterson, J. (1909), 'Report on Scottish Ornithology for 1908', *Ann. Scot. Nat. Hist.*, 1909, p. 198. (Robins feeding fledgling blackbird.)

p. 97 Corbin, G. B. (1875), 'Young cuckoo and robin: a case of adoption', *Zool.*, 2, 10: 4695-6. (Taken over from meadow pipit.)

p. 98 Eckermann, J. P. (1836), *Gespräche mit Goethe*. English translation by J. Oxenford in Everyman edition, 1930, p. 243.

p. 98 Steinbacher, J. (1936), 'Zur Frage der Geschlechtsreife von Kleinvögeln', *Beitr. z. Fortpflanz. biol. d. Vög.*, 12: 139-44. (Young of first brood helping to feed second brood.)

p. 98 Skutch, A. (1935), 'Helpers at the nest', *Auk*, 52: 257-73. (Young of first brood helping to feed second brood.)

p. 99 Renier, G. J. (1934), *A Tale of Two Robins*.

p. 99 Kirkman, F. B. (1937), *Bird Behaviour*. (Recognition of eggs and young by black-headed gull.)

p. 100 Goethe, F. (1937), see reference given in full under Ch. III, p. 50. (Recognition of eggs and young by herring-gull.)

p. 101 Rand, A. L. (1941), 'Development and enemy recognition of the curve-billed thrasher *Toxostoma curvirostre*', *Bull. Amer. Mus. Nat. Hist.*, 78: 213-42.

p. 101 Osmaston, B. B. (1934), 'Robin with four broods', *Brit. Birds*, 28: 113-16.

p. 101 Owen, J. H. (1914), 'Short interval between two nests of robin', *Brit. Birds*, 8: 50. (Four successive broods.)

p. 102 Fisher, J. (1939), *Birds as Animals*, p. 194. (Summarizes cases of female bird leaving fledglings to start new brood.)

CHAPTER 8

p. 104 Colquhoun, M. K. (1940), 'A note on the territorial behaviour of robins in cold weather', *Brit. Birds*, 33: 274-5. (Ringed male returning next spring.)

p. 105 Reeks, H. (1864), 'Scarcity of robins during winter: Does the female migrate?', *Zool.*, 22: 8949.

p. 109 Meinertzhagen, R. (1933), *Bull. Brit. Orn. Club*, 54: 8 (also private letter on sex of shot bird). (Robins on Ushant.)

p. 109 Olivier, G. (1938), 'Les Oiseaux de la Haute-Normandie', *L'Oiseau*, N. S., 8: 207. (Robin in Normandy.)

p. 109 Ticehurst, N. F. (1938), in Witherby, H. F., *et al*, *The Handbook of British Birds*, vol. II, p. 203. (General account of robin's migration.)

p. 109 Barrington, R. M. (1900), *The Migration of Birds at Irish Light Stations*, p. 399 *et al*.

p. 109 Lockley, R. M., *in litt*. (Ringed robins on Skokholm.)

p. 109 Patten, C. J. (1913), 'Robins on migration observed at the Tuskar rock and lighthouse', *Zool.*, 4, 17: 2-14.

p. 110 Payn, W. A. (1934), 'Migration of robins', *Brit. Birds*, 27: 230-1.

p. 110 Eds. (1909), 'Rare birds on the Isle of May (Firth of Forth)', *Brit. Birds*, 2: 346.

p. 112 Michener, H. and J. R. (1935), 'Mocking-birds, their territories and individualities', *Condor*, 37: 97-140.

p. 112 Schäfer, E., and de Schaunsee, R. M. (1939), 'Zoological results of the Second Dolan expedition to Western China and Eastern Tibet 1934-6. Part II, Birds', *Proc. Acad. Nat. Sci. Philadelphia*, 90: 223. (Güldenstadt's redstart wintering in Tibet.)

p. 112 Drost, R. (1935), 'Ueber das Zahlenverhältnis von Alter und Geschlecht auf dem Herbst und Frühjahrszuge', *Vogelzug*, 6: 177-82. (Proportion of males, females, and juveniles on migration.)

p. 112 Weimann, R. (1938), 'Beringungsergebnisse schlesischer und sächsischer Amseln (*Turdus merula merula* [L.])', *Ber. Ver. Schles. Orn.*, 23: 1-14. (Proportion of male, female, and juvenile blackbirds which migrate.)

p. 112 Niethammer, G. (1937), *Hanbuch der deutschen Vögelkunde*. (Relation of sex and age to migration in chaffinch, robin, etc., in Germany.)

p. 112 Thomson, A. L. (1939), 'The migration of the gannet: results of marking in the British Isles', *Brit. Birds*, 32: 282-9.

p. 112 Lack, D. (1943-4), 'The problem of partial migration', *Brit. Birds*,

37: 122-50; 37: 122-30, 143-50. (Relation of sex and age to migration, etc.)

p. 113 Alexander, C. J. (1917), 'Observations on birds singing in their winter quarters and on migration', *Brit. Birds*, 11: 98-102. (Robins singing in winter in Italy.)

p. 113 Heinroth, O., private letter on robins singing in winter quarters in Dalmatia.

p. 113 Rooke, K. B. (1947), 'Notes on robins wintering in Northern Algeria', *Ibis*, 89: 204-10. (Also quotes Salerno observation.)

p. 113 Blyth, E. (1833), 'New facts on the migrations of various birds and insects', Rennie's *Field Naturalist* for November, 1833: 466-72. (Robins on ship.)

p. 114 Rowan, W. (1929), 'Experiments in bird migration', *Proc. Boston Nat. Hist. Soc.*, 39: 151-208. Also (1931), *The Riddle of Migration.*

p. 114 Wolfson, A. (1942), 'Regulation of spring migration in juncos', *Condor*, 44: 237-63.

p. 114 Schildmacher, H. (1938), 'Zur Physiologie des Zugtriebes IV, Weitere Versuche mit künstlich veränderter Belichtungszeit', *Vogelzug*, 9: 146-52. (Night restlessness.)

p. 114 Palmgren, P. (1937), 'Auslösung der Frühlingsunruhe durch Wärme bei gekäfigten Rotkehlchen, *Erithacus rubecula* (L.)', *Ornis Fennica*, 14: 71-3.

p. 115 Putzig, P. (1938), 'Beobachtungen über Zugunruhe beim Rotkehlchen (*Erithacus rubecula*)', *Vogelzug*, 9: 10-14.

p. 115 Barrington, Hon. Daines (1771), 'An essay on the periodical appearing and disappearing of certain birds, at different times of year', *Phil. Trans. Roy. Soc.*, 62: 265-326. (Best statement of case for hibernation of swallow.)

p. 115 Boswell, J. (1791), *Life of Johnson* (in Oxford, 1768; swallows conglobulating).

CHAPTER 9

p. 119 Burkitt, J. P. (1938), 'Eleven-year-old robin', *Irish Nat. Journ.*, 7: 85.

p. 119 Drost, R., and Schüz, E. (1932), 'Vom Zug des Rotkehlchens, *Erithacus r. rubecula* [L.]', *Vogelzug*, 3: 164-9. (Heligoland robin in eleventh year.)

p. 120 Five papers on age of birds:
  (i) Gurney, J. H. (1899), 'On the comparative ages to which birds live', *Ibis*, 19-42.
  (ii) Mitchell, P. C. (1911), 'On longevity and relative viability in mammals and birds; with a note on the theory of longevity', *Proc. Zool. Soc. Lond.*, 425-548.
  (iii) Flower, S. S. (1925), 'Contributions to our knowledge of the

duration of life in vertebrate animals. IV. Birds', *Proc. Zool. Soc. Lond.*, 1366-422.

(iv) Witherby, H. F. (1926), 'The duration of life in birds', *Brit. Birds*, 20: 71-3.

(v) Nice, M. M. (1934), 'Ten-year-old passerines', *Condor*, 36: 245.

p. 120 Heilbrunn, L. V. (1938), *An Outline of General Physiology*, pp. 513-19. (Age of dog, cat, mouse, and many other animals.)

p. 120 Clarke, A. J. (1927), *Comparative Physiology of the Heart*, p. 146. (Heart-beat of redstart.)

p. 120 Wallace, A. R. (1889), *Darwinism*, pp. 25-6.

p. 122 Lack, D. (1943), 'The age of the blackbird', *Brit. Birds*, 36: 166-75, and (1943), 'The age of some more British birds', *Brit. Birds*, 36: 193-221.

p. 123 Ruiter, C. J. S. (1941), 'Waarnemingen omtrent de levenswijze van de Gekraagde Roodstaart, *Phoenicurus ph. phoenicurus* (L.)', *Ardea*, 30: 175-214 (age of redstart).

p. 124 Dublin, L. I., and Lotka, A. J. (1936), *Length of Life: a study of the life-table*, pp. 14-17. (Human survival.)

p. 125 Pearl, R. (1920), *The Biology of Death*. (Survival of humans and laboratory animals.)

p. 125 Macdonell, W. R. (1913), 'On the expectation of life in ancient Rome, and in the provinces of Hispania and Lusitania, in Africa', *Biometrika*, 9: 366-80.

p. 126 Leslie, P. H., and Ranson, R. H. (1940), 'The mortality, fertility and rate of natural increase of the vole (*Microtus agrestis*) as observed in the laboratory', *J. Animal Ecol.*, 9: 27-52.

p. 128 Lack, D. (1946) and (1948), 'Clutch and brood-size in the robin', *Brit. Birds*, 39: 98-109, 130-135; 41: 98-104, 130-7.

CHAPTER 10

p. 132 Jourdain, F. C. R. (1938), in *A Handbook of British Birds*, Witherby, H. F., *et al.*, vol. II, p. 203. (Food of robin.)

p. 132 Steinfatt, O. (1937), 'Nestbeobachtungen beim Rotkehlchen (*Erithacus r. rubecula*), etc.', *Verh. Ornith. Ges. Bayern*, 21: 139-145. (Food of robin.)

p. 132 Lebeurier, E., and Rapine, J. (1936), 'Ornithologie de la Basse-Bretagne', *L'Oiseau*, n.s., 6: 252-71. (Food of robin.)

p. 132 Wordsworth, W. (nineteenth century), 'Redbreast chasing the butterfly'.

p. 132 Dunlop, E. B., unpublished note in MS. diary in Edward Grey Institute of Field Ornithology, Oxford, and G. Hale Carpenter private letter. (Robin rejecting white and small tortoiseshell butterflies).

p. 132 Collinge, W. E. (1924-7), *Food of Some British Wild Birds*, 2nd edition, pp. 150-2.

p. 133 Ticehurst, C. B. (1932), *A History of the Birds of Suffolk*, p. 162. (Robin attending on moles working.)

p. 133 Hudson, R. (1928), 'Robin taking fish from tank', *Brit. Birds*, 21: 260.

p. 133 Conway, C. (1835), 'Sketches of the natural history of my neighbourhood. No. 3. Fragments of ornithology', *Loudon's Mag. Nat. Hist.*, 8: 545-6. (Robin taking something off surface of stream.)

p. 133 Coward, T. A. (1933), *The Birds of the British Isles and their Eggs*, 1st ser., 4th ed., p. 216. (Robin picking flies off surface of pool.)

p. 133 Morris, F. O. (1853), *A History of British Birds*, vol. III, p. 120. (Robin taking object off water.)

p. 133 Selous, E. (1901), *Bird Watching*, pp. 229-30. (Robin feeding from scum on a stream.)

p. 133 Ingram, G. C. S., and Salmon, H. M. (1934), *Birds in Britain To-day*, p. 36. (Robin catching minnows.)

p. 133 Oldham, C. (n.d.), In MS. notes of F. C. R. Jourdain at Edward Grey Institute of Field Ornithology, Oxford. (Robin pellet.)

p. 134 Tucker, B. W. (1944), 'The ejection of pellets by passerine and other birds', *Brit. Birds*, 38: 50-2.

p. 134 Crewe, H. (1857), 'Note on the robin (*Sylvia rubecula*) and Butcher Bird (*Lanius collurio*)', *Zool.*, 15: 5516. (Ejecting pellets.)

p. 134 Brooks-King, M. (1944), 'Some observations on a tame robin', *Brit. Birds*, 38: 130-2. (Tame robins eating butter, margarine, floor polish, sulphate of potash.)

p. 134 Dale, F., private letter on tame robin eating butter and refusing margarine.

p. 134 Owen, J. H., private letter on robin eating meat and carrion.

p. 134 St John, C. (1882), *Natural History and Sport in Moray*, p. 110. (Robin eating meat.)

p. 134 Jourdain, F. C. R., and Witherby, H. F. (1918), 'The effect of the winter of 1916-17 on our resident birds', *Brit. Birds*, 11: 266-71 and 12: 26-35. (See also H. M. Wallis, 10: 267-8 and C. J. Carroll, 11: 26-8.)

p. 134 Jourdain, F. C. R., and Witherby, H. F. (1929), 'Report on the effect of severe weather in 1929 on bird life', *Brit. Birds*, 23: 154-8.

p. 134 Ticehurst, N. F., and Witherby, H. F. (1940), 'Report on the effect of severe weather of 1939-40 on bird life in the British Isles', *Brit. Birds*, 34: 118-32, 142-55.

p. 135 Lebeurier, E., and Rapine, J. (1936), 'Ornithologie de la Basse-Bretagne', *L'Oiseau*, n.s., 6: 252-71. (Weight of French robins.)

p. 135 Harrison, J. M., and Hurrell, H. J., private letters on finding emaciated robins in hard weather, Hurrell also noting abandonment of territories in Devon in severe weather.

p. 135 D'Urban, W. S. M., and Mathew, M. A. (1895), *The Birds of Devon*, p. 15. (Robins entering house in cold weather.)

p. 135 Colquhoun, M. K. (1940), 'A note on the territorial behaviour of robins in cold weather', *Brit. Birds*, 33: 274-5.

p. 136 Ranson, J. (1863), 'The "Query about the robin"', *Zool.*, 21: 8843-4. (Cat eating robin.)

p. 136 Wintle, G. S. (1864), 'Notes about robins', *Zool.*, 22: 8877-8. (Fox eating robin.)

p. 136 Dunsheath, M. H., and Doncaster, C. C. (1941), some observations on roosting birds, *Brit. Birds*, 35: 144-5.

p. 136 Witherby, H. F. (1932), 'Curious fatality to a redbreast and its young', *Brit. Birds*, 26: 96. (Robin with horsehair in gizzard.)

p. 136 Gurney, J. H. (1874), 'Accident to a weasel and to a redbreast', *Zool.*, 2, 9: 4194-5. (Robin caught between slates.)

p. 136 Long, F. (1883), 'Singular accident to a robin', *Zool.*, 3, 7: 123-124. (Beak embedded in neck.)

p. 137 Cuvier, Baron G. (1817), *Le Règne animal distribué d'après son organisation*.

p. 137 Waterton, C. (1884), *Essays on Natural History, chiefly Ornithology*, 2nd series, p. 99.

p. 138 Gurney, J. H. (1871), 'On the ornithology of Algeria', *Ibis*: 82. (Robins in market.)

p. 138 Godard, A. (1916), 'Les Jardins-Volières', *Rev. Franc. d'Ornith.*, 4: 233-9.

p. 138 Newton, A. (1893-6), *Dictionary of Birds*, pp. 771-2. (Robins killed for Christmas cards and trimmings.)

p. 140 Alexander, W. B. (1945), 'The index of heron population, 1944', *Brit. Birds*, 38: 232-4.

CHAPTER II

p. 143 Gesner, Conrad (1555), *Historiæ Animalium* III *De Avium Naturæ*. p. 697.

p. 143 Olina, G. P. (1622), *Uccelliera*, pp. 1, 16.

p. 143 Buffon, G., Count de (1771-83), *Histoire Naturelle des Oiseaux*. (See note under Ch. I, p. 216.)

p. 146 Howard, H. E. (1907-14), *A History of the British Warblers*, and (1920), *Territory in Bird Life*.

p. 146 Aristotle, *Historia Animalium*, trans. D'A. W. Thompson, 1910.

p. 146 Ticehurst, N. F. (1934), 'Letter to the editors', *Brit. Birds*, 27: 308. (Territory in mute swan.)

p. 147 White, G. (1789), Letter XI to Barrington, in *The Natural History and Antiquities of Selborne*.

p. 147 Goldsmith, O. (1774), *An History of the Earth and Animated Nature*, 5: 301.

p. 147 Montagu, G. (1802), 'Introduction', in *Ornithological Dictionary*, pp. xxviii-xxxiii.

p. 147 Altum, B. (1868), *Der Vogel und sein Leben*. See also Mayr, E. (1935), 'Bernard Altum and the territory theory', *Proc. Linn. Soc. New York*, 45-6: 24-38.

p. 147 Five works discussing the value of territory:
  (i) Nicholson, E. M. (1927), *How Birds Live*.
  (ii) Lack, D. and L. (1933), 'Territory reviewed', *Brit. Birds*, 27: 179-99.
  (iii) Tinbergen, N. (1936), 'The function of sexual fighting in birds, and the problem of the origin of "territory"', *Bird Banding*, 7: 1-8.
  (iv) Tinbergen, N. (1939), 'The behaviour of the snow-bunting in spring', *Trans. Linn. Soc. New York*, 5: 1-94.
  (v) Nice, M. M. (1941), 'The rôle of territory in bird life', *Amer. Midland Nat.*, 26: 441-87.

p. 150 Siivonen, L. (1939), 'Zur Oekologie und Verbreitung der Singdrossel (*Turdus ericetorum philomelos* Brehm)', *Ann. Zool. Soc. Zool.-Bot. Fenn. Vanamo*, 7 (i): 1-289. (Abstra. *Bird Banding*, 1940, 11: 28.) (Size of territory in song-thrush.)

p. 150 Erickson, M. M. (1938), 'Territory, annual cycle, and numbers in a population of wren-tits (*Chamæa fasciata*)', *Univ. Calif. Publ. Zool.*, 42: 247-334. (Size of territory.)

p. 151 Huxley, J.(1934), 'A natural experiment on the territorial instinct', *Brit. Birds*, 27: 270-7. (Limit to compression of territory.)

p. 153 Lack, D. (1935), 'Territory and polygamy in a bishop-bird (*Euplectes hordeacea hordeacea* [Linn.])', *Ibis*, 13 (5): 817-36.

p. 154 Renier, G. J. (1934), *A Tale of Two Robins*. (Abandonment of territory in cold weather.)

p. 154 Colquhoun, M. K. (1940), 'A note on the territorial behaviour of robins during cold weather', *Brit. Birds*, 33: 274-5.

p. 154 Morley, A. (1941), 'The behaviour of a group of resident British starlings (*Sturnus v. vulgaris* Linn.) from October to March', *The Naturalist*, 788: 55-61.

p. 154 Lack, D. (1939), 'The display of the blackcock', *Brit. Birds*, 32: 290-303.

p. 155 Brewster, W. (1898), 'Revival of the sexual passion of birds in autumn', *Auk*, 15: 194-5.

p. 155 Morley, A. (1943), 'Sexual behaviour in British birds between October and January', *Ibis*, 85: 132-58.

p. 155 Browne, M. (1885), 'Notes on the vertebrate animals of Leicestershire', *Zool.*, 3, 9: 337. (Robin breeding in autumn.)

p. 155 Bird, G. (1932), 'Late nests of hedge-sparrow and robin in Suffolk', *Brit. Birds*, 26: 225.

p. 152 Leonard, S. L. (1939), 'Induction of singing in female canaries by injections of male hormone', *Proc. Soc. Exp. Biol. and Med.*, 41: 229-30.

p. 155 Shoemaker, H. H. (1939), 'Effect of testosterone propionate on behaviour of the female canary', *Proc. Soc. Exp. Biol. and Med.*, 41: 299-302.

p. 155 Bullough, W. S., and Carrick, R. (1940), 'Male behaviour of the female starling (*Sturnus v. vulgaris*) in autumn', *Nature*, 145: 629.

p. 156 Bullough, W. S. (1943), 'Autumn sexual behaviour and the resident habit of British birds', *Nature*, 151: 531.

p. 156 Michener, H. and J. R. (1935), 'Mocking-birds, their territories and individualities', *Condor*, 37: 97-140.

p. 157 Lack, D. (1944), 'The problem of partial migration', *Brit. Birds*, 37: 144-5. (General discussion of autumn sexual behaviour and migration.)

CHAPTER 12

p. 159 Thompson, E. P. (1845), *The Note-book of a Naturalist*, pp. 72-5. (Robin attacking stuffed specimen.)

p. 159 Morris, F. O. (1853), *A History of British Birds*, vol. III, pp. 111-13. (Two records of robin attacking stuffed specimen.)

p. 159 Allen, A. A. (1934), 'Sex rhythm in the ruffed grouse (*Bonasa umbellus* L.) and other birds', *Auk*, 51: 180-99. (Use of stuffed specimens.)

p. 159 Chapman, F. M. (1935), 'The courtship of Gould's manakin (*Manacus vitellinus vitellinus*) on Barro Colorado Island, Canal Zone', *Bull. Amer. Mus. Nat. Hist.*, 68: 471-525. (Use of stuffed specimen.)

p. 163 Congreve, W. M. (1924), 'Cannibalistic propensity of a redbreast', *Brit. Birds*, 17: 251.

p. 166 Howard, H. E. (1935), *The Nature of a Bird's World*. (Incidents with moorhen, yellow-hammer, linnet and reed-bunting.)

p. 166 Shakespeare, W. (1600), *A Midsummer Night's Dream*, Act III, Sc. i.

p. 168 Watson, J. B. (1908), 'The behaviour of noddy and sooty terns', *Papers Tortugas Lab. Carnegie Inst.*, 103: 187-255. Certain parallels with this work on robins occur in: Pelkwyk, J. T., and Tinbergen, N. (1937), 'Eineeiz biologische Analyse einiger Verhaltensweisen von *Gasterosteus aculeatus* L.', *Zeitschr. f. Tierpsychol.*, 1: 193-200.

CHAPTER 13

p. 171 Clark, J., and Rodd, F. R. (1906), 'The Birds of Scilly', *Zool.*, 4, 10: 244. (Robin attacking redstart.)

p. 172 Rooke, K. B. (1947), 'Notes on robins wintering in Northern Algeria', *Ibis*, 89: 204-10. (Robin attacking black redstart.)

p. 172 Heinroth, O., private letter on robin attacking common redstart.

p. 172 Palmer, A. H. (1895), *The Life of Joseph Wolf*, p. 145. (Robin attacking stuffed common redstart.)

p. 172 Stout, G. F. (1901), *Manual of Psychology.* (Quotation on language in relation to animal minds, also quoted by H. E. Howard (1920), *Territory in Bird Life*).

p. 173 Lorenz, K. (1935, 1937, 1939), see full references under Ch. III, p. 50 (i).

p. 173 Tinbergen, N. (1939), 'On the analysis of social organization among vertebrates, with special reference to birds', *Amer. Midland Nat.*, 21: 210-34. (Substitutes the term 'signal' for 'releaser'.)

p. 173 Goethe, F. (1937), see full reference under Ch. III, p. 50. (Herring-gull's recognition of parents.)

p. 177 Teager, C. W. (1939), 'Display of robins', *Countryside*, 495-6. (Continuance of threat display between pair.)

p. 177 Lindsay-Blee, M. (1939), 'Display of robins', *Countryside*, 510. (Continuance of threat display between pair.)

p. 178 Leonard, S. L. (1939), 'Induction of singing in female canaries by injections of male hormone', *Proc. Soc. Exp. Biol. and Med.*, 41: 229-30.

p. 178 Shoemaker, H. H. (1939), 'Effect of testosterone propionate on behaviour of female canary', *Proc. Soc. Exp. Biol. and Med.*, 41: 299-302.

p. 178 Noble, G. K., and Wurm, M. (1938), 'Effect of testosterone pro-pionate on the black-crowned night heron', *Anat. Rec.*, 72, no. 4 suppl.: 60.

CHAPTER 14

p. 182 Gurney, J. H. (1921), *Early Annals of Ornithology*, p. 154, quoting Ælian, *De Animalium Natura*, II, second century AD. (Kite taking human hair.)

p. 182 Slater, H. H. (1883), 'Field notes in Norway in 1881', *Zool.*, 3, 7: 7. (Robin very shy.)

p. 182 Ticehurst, C. B., and Whistler, H. (1925), 'A contribution to the ornithology of Navarre, Northern Spain', *Ibis*: 455. (Robin shy.)

p. 182 Bannerman, D. A. (Unpublished MS.), 'Birds of the Canary Islands' (Deposited in Brit. Mus. Nat. Hist.). (Robin shy.)

p. 182 Alexander, H. G. (1924), 'Birds of the Lake of Geneva', *Ibis*: 281. (Robin replaced by black redstart.)

p. 182 Wallis, H. M. (1895), 'Notes on the birds of the Central Pyrenees', *Ibis*: 67. (Robin replaced by black redstart.)

p. 182 Guériot, A. (1918), 'Les familiarités du rouge-gorge', *Rev. Franc. d'Ornith.*, 5: 249-55. (Tame French robin.)

p. 182 Quentin, J. (1928), 'Note sur la familiarité et le commensalisme du Rouge-Gorge. *Erithacus r. rubecula* (L.)', *Rev. Franc. d'Ornith.*, 12: 257-8. (Tame French robin.)

p. 182 Tucker, B. W. (1938), partly quoting E. V. Baxter and L. J. Rintoul in Witherby, H. F., *et al*, *The Handbook of British Birds*, vol. II, p. 199.

p. 182 Meinertzhagen, R. (1933), *Bull. Brit. Orn. Club*, 54: 8. (Robins on Ushant.)

p. 183 Grey of Fallodon, Viscount (1927), *The Charm of Birds*, pp. 191-201.

p. 184 Nice, M. M. (1943), '*Studies in the life history of the Song-Sparrow*. II. The behaviour of the song-sparrow and other passerine birds', *Trans. Linn. Soc. New York*, 6: 7.

p. 184 Brooks-King, M. (1944), 'Some observations on a tame robin', *Brit. Birds*, 38: 130-2. (Three baths daily.)

p. 185 Morley, A. (1942), 'Effects of baiting on the marsh-tit', *Brit. Birds*, 35: 261-6. (Greeting benefactor with calls.)

p. 185 Grey, Viscount (1925), quoted by J. H. Burkitt, *Brit. Birds*, 18: 256.

p. 185 Nicholas, W. W. (1942), 'A year in the life of the robin', *Field*, 179: 580. (Robins buffeting human being.)

p. 185 Witherby, H. F. (1920), 'Unusual boldness of robin in defence of young', *Brit. Birds*, 14: 92.

p. 185 Jourdain, F. C. R., MS. diaries deposited at Edward Grey Institute of Field Ornithology. (Robin attacking man.)

p. 185 Heinroth, O. and M. (1924-6), *Die Vögel Mitteleuropas*, vol. I, pp. 10-14.

p. 185 Renier, G. J. (1934), *A Tale of Two Robins*.

p. 186 Walton, Izaak (1653), *The Compleat Angler*.

p. 186 Whitman, C. O. (1919), *The Behaviour of Pigeons*, posthumous works of Charles Otis Whitman, vol. III, edit. H. A. Carr.

p. 187 Lorenz, K. (1935, 1937, and 1939), see full references under Ch. III, p. 50 (i). (The account of Portielje's bittern is in the 1937 paper.)

p. 188 Wormald, H. (1909), 'A tame snipe and its habits', *Brit. Birds*, 2: 249-58.

p. 189 Darling, F. F. (1940), *Island Years*, pp. 65-7. (Transference of behaviour in young grey-lag goose.)

p. 189 Romanes, G. J. (1883), *Mental Evolution in Animals*, pp. 183-4. (Wigeon attached to peacock and Chinese goose to dog.)

p. 189 Thorpe, W. H. (1938), 'Further experiments on olfactory conditioning in a parasitic insect. The nature of the conditioning process', *Proc. Roy. Soc.*, B, 126: 370-97. Also (1939), 'Further studies in pre-imaginal olfactory conditioning in insects', *Proc. Roy. Soc.*, B, 127: 424-33.

CHAPTER 15

p. 192 Howard, H. E. (1935), *The Nature of a Bird's World*.

p. 193 Fabre, J. H. (1911), *Souvenirs entomologiques*, 6th series, 6th edition, pp. 332-52. Quoted by Russell, E. S., (1938), *Animal Behaviour*.

p. 194 Boswell, J. (1791), *Life of Johnson*. (In Oxford, 1768.)

p. 194 Cuvier, Baron G. (1817), *La Règne animal distribué d'après son organisation*.

p. 194 Darwin, E. (1801), *Zoonomia*.

p. 195 Lorenz, K. (1935, 1937, and 1939), full references under Ch. III, p. 50 (i).

p. 196 Craig, W. (1918), 'Appetites and aversions as constituents of instincts', *Biol. Bull.*, 34: 91.

p. 196 Katz, D. (1937), *Animals and Men: Studies in Comparative Psychology*. (Quoting Bethe's experiments on dogs with only two limbs).

p. 196 Nice, M. M., and Pelkwyk, J. T. (1941), 'Enemy recognition by the song-sparrow', *Auk*, 58: 197-214.

p. 196 Lack, D. (1941), 'Some aspects of instinctive behaviour and display in birds', *Ibis*: 407-41. (Gives references to much discussed in this chapter.)

p. 196 Taibell, A. (1928), 'Risveglio artificiale de instinti tipicamente f emminili nei maschi di taluni ucceli', *Atti della Societa die Naturalisti e Matematici di Modena*, ser. 9, vol. 7 (59), 93-102. (Turkey cock tied down to eggs.) (I have read only an abstract prepared by B. Roberts.)

p. 196 Morgan, C. Lloyd (1896), *Habit and Instinct*.

p. 199 Roberts, B. B. (1934), 'Notes on the birds of Central and South-East Iceland, with special reference to food habits', *Ibis*: 252 (Behaviour of eiders.)

p. 201 Bacon, Francis (1622-3), *Historia Vitæ et Mortis*.

CHAPTER 16

p. 202 Lack, D. (1948), 'Notes on the ecology of the robin', *Ibis*: 90: 252-79. (Habitat, food, winter habits.)

p. 207 Lack, D., and Silva, E. (1949), 'The weight of nestling robins', *Ibis*: 91: 64-78. (Also feeding frequencies.)

p. 209 Burton, R. E. (1947), 'Robins rearing own young and cuckoo in same nest', *Brit. Birds*, 40: 149-50.

p. 212 Snow, D. W. (1958), *A Study of Blackbirds*.

p. 215 Friedmann, H. (1955), *The Honey-Guides*. (*U.S. Nat. Mus. Bull.*, 208.)

p. 216 Lack, D. (1954), 'Two robin populations', *Bird Study*, 1: 14-17.

p. 216 Beven, S. (1963), 'Population changes in a Surrey Oakwood during fifteen years', *Brit. Birds*, 56: 307-23.

p. 218 Huxley, J. (1934), 'A natural experiment on the territorial instinct', *Brit. Birds*, 27: 270-7.

NOTES TO
IN DAVID LACK'S FOOTSTEPS

1 Lack, D. (1953), *Life of the Robin*, p. 17.

2 Harper, D. G. C. (1985), 'Brood division in Robins', *Animal Behaviour*, 33: 466-480.

3 Madsen, V. (1997), 'Sex determination of Continental European Robins *Erithacus r. rubecula*', *Bird Study*, 44: 239-44; Campos, A. R., Catry, P., Ramos, J., and Robalo, J. I. (2011), 'Competition among European Robins *Erithacus rubecula* in the winter quarters: Sex is the best predictor of priority to experimental food resources', *Ornis Fennica*, 88: 226-233.

4 Snow, D. W. (1969), 'The moult of British thrushes and chats', *Bird Study*, 16: 115-29.

5 Ginn, H. B., and Melville, D. S. (1993), *Moult in Birds*.

6 Hoelzel, A. R. (1986), 'Song characteristics and response to playback of male and female Robins *Erithacus rubecula*', *Ibis*, 128: 115-27.

7 Brindley, E. L. (1991), 'Response of European Robins to playback of song: neighbour recognition and overlapping', *Animal Behaviour*, 41: 503-12.

8 Bremond, J.-C. (1968), 'Recherches sur la sémantique et les éléments dans les signaux acoustiques du Rouge-gorge (*Erithacus rubecula*)', *Terre et Vie*, 22: 109-220.

9 Thomas, R. J., Cuthill, I. C., Goldsmith, A. R., Cosgrove, D. F., Lidgate, H. C., and Burdett Proctor, S. L. (2003), 'The trade-off between singing and mass gain in a daytime-singing bird, the European Robin', *Behaviour*, 140: 387-404.

10 Dabelsteen, T., McGregor, P. K., Holland, J., Tobias, J. A., and Pedersen, S. B. (1997), 'The signal function of overlapping singing in male Robins', *Animal Behaviour*, 53: 249-56.

11 Kempanaers, B., Borgström, P., Loës, P., Schlicht, E., and Valcu, M. (2010), 'Artificial night lighting affects dawn song, extra-pairing success, and lay date in songbirds', *Current Biology*, 20: 1735-9.

12 Fuller, R. A., Warren, P. H., and Gaston, P. K. (2007), 'Daytime noise predicts nocturnal singing in urban Robins', *Biology Letters*, 3: 368-70.

13 McLaughlin, K. E., and Kunc, H. P. (2013), 'Experimentally induced noise levels change spatial and singing behaviour', *Biology Letters*, 9: 20120771; Montague, M. J., Danek-Gontard, M., and Kunc, H. P. (2013), 'Phenotypic plasticity affects the response of a sexually selected trait to anthropogenic noise', *Behavioural Ecology*, 24: 343-8.

14 McMullen, H., Schmidt, R., and Kunc, H. P. (2013), 'Anthropogenic noise affects vocal interactions', *Behavioural Processes*, 103: 125-8.

15 Jovani, R., Aviles, J. M., and Rodriguez-Sanchez, F. (2011), 'Age-related sexual plumage dimorphism and badge framing in the European robin *Erithacus rubecula*', *Ibis*, 154: 147-54.

16 Tobias, J. (1997), 'Asymmetric territorial contests in the European robin: the role of settlement costs', *Animal Behaviour*, 54: 432-6.

17 e.g.: Ginter, M., Rosińska, K., and Remisiewicz, M. (2005), 'Variation in the extent of greater wing coverts moult in Robins (*Erithacus rubecula*) migrating through the Polish Baltic coast', *Ring*, 27: 177-87.

18 Jenni, L., and Winkler, R. (2011), *Moult and Ageing of European Passerines*.

19 Dunn, M., Copelston, M., and Workman, L. (2004), 'Trade-offs and seasonal variation in territorial defence and predator evasion in the European Robin *Erithacus rubecula*', *Ibis*, 146: 77-84.

20 e.g.:. Harper, D. G. C. (1985), 'How do male Robins *Erithacus rubecula* discover that their chicks have hatched?', *Ibis*, 127: 262-6.

21 Harper, D. G. C. (1985), 'Pairing strategies and mate choice in female Robins *Erithacus rubecula*', *Animal Behaviour*, 33: 862-75.

22 des Forges, G. (1959), 'Two Robins lay in one nest', *British Birds*, 52: 390.

23 e.g.: Watson, J. (2010), *The Golden Eagle*.

24 e.g.: Galbraith, J. A., Beggs, J., Jones, D. N., and Stanley, M. C. (2015), 'Supplementary feeding restructures bird communities', *Proceedings of the National Academy of Sciences*, 112: e2648-57.

25 East, M. (1981), 'Aspects of courtship and parental care of the European Robin *Erithacus rubecula*', *Ornis Scandinavica*, 12: 230-9.

26 Tobias, J., and Seddon, N. (2002), 'Female begging in European robins: Do neighbours eavesdrop for extrapair copulations?', *Behavioral Ecology*, 13: 637-42.

27 Harper, D. G. C. (1984), 'Economics of foraging and territoriality in the European Robin *Erithacus rubecula*', PhD thesis, University of Cambridge.

28 Ferguson-Lees, J., Castell, R., and Leech, D. (2011), *A Field Guide to Monitoring Nests*.

29 Reviewed by: Barve, S., and Mason, N. A. (2015), 'Interspecific competition affects evolutionary links between cavity nesting, migration and clutch size in Old World Flycatchers (*Muscicapidæ*)', *Ibis*, 157: 299-311.

30 Jensen, R. A. C., and Jensen, M. K. (1970), 'First breeding records of the Herero Chat *Numibornis herero*, and taxonomic implications', *Ostrich*, 8, suppl.: 105-6.

31 East, M. (1981), 'Alarm calling and parental investment in the Robin
    Erithacus rubecula', *Ibis*, 123: 223-30.

32 Fennessy, G., and Harper, D. (2002), 'European Robin *Erithacus
    rubecula*', in Wernham, C. V., Toms, M. P., Marchant, J. M., Clark,
    J. A., Siriwardena, G. M., and Baillie, S. R. (eds), *The Migration
    Atlas of Birds in Britain and Ireland*, pp. 498-501.

33 Biebach, H. (1983), 'Genetic determination of partial migration in
    the European Robin *Erithacus rubecula*', *Auk*, 100: 601-6.

34 Harper, D. G. C. (2003), 'The vanishing Robin mystery', *Bird
    Watching*, 17 (1): 7-11.

35 Tellería, J. L. (2014), 'Has the number of European Robins *Erithacus
    rubecula* wintering in Spain decreased?', *Ardeola*, 61: 389-91.

36 Tellería, J. L. (2015), 'Retraction', *Ardeola*, 62: 185-6.

37 Dominguez, M., Barba, E., Carlo, J. L., Lopez, G. M., and Monrós,
    J. S. (2007), 'Seasonal interchange of the European Robin *Eritha-
    cus rubecula* populations in an evergreen holm oak forest', *Acta
    Ornithologica*, 42: 15-21.

38 Campos, A. R., Catry, P., Tenreiro, P., Neto, J. M., Pereira, A. C.,
    Brito, R., Cardoso, H., Ramos, J. A., Bearhop, S., and Newton, J.
    (2011), 'How do Robins *Erithacus rubecula* resident in Iberia
    respond to seasonal flooding by conspecific migrants?', *Bird
    Study*, 58: 435-42.

39 Catry, P., Campos, A., Almada, V., and Cresswell, W. (2004),
    'Winter segregation of migrant European Robins *Erithacus
    rubecula* in relation to sex, age and size', *Journal of Avian
    Biology*, 35: 204-9.

40 Campos, A. R., Catry, P., Ramos, J., and Robalo, J. I. (2011),
    'Competition among European Robins *Erithacus rubecula* in
    the winter quarters: Sex is the best predictor of priority to
    experimental food resources', *Ornis Fennica*, 88: 226-33.

41 Merkel, F. W., and Wiltschko, W. (1965), 'Magnetismus und
    Richtungsfinden zugunruhiger Rotkelchen *Erithacus rubecula*',
    Vogelwarte, 23: 71-3; Wiltschko, W. (1968), 'Uber den einfluss
    statischer magnetfelder auf die zugorientierung der Rotkelchen
    Erithacus rubecula', *Zeitschrift für Tierpsychologie*, 25: 536-8.

42 Wiltschko, W., and Wiltschko, R. (1972), 'Magnetic compass of
    European robins', *Science*, 176: 62-4.

43 Recent reviews: Begall, S., Burda, H., and Malkemper, E. P. (2014),
    'Magnetoreception in mammals', *Advances in the Study of
    Behavior*, 46: 45-87; Heinrich, B. (2015), *The Homing Instinct:
    Meaning & Mystery in Animal Migration*.

44 Reviewed by: Stevens, M. (2013), *Sensory Ecology, Behaviour &
    Evolution*.

45 Wiltschko, R., and Wiltschko, W. (2014), 'Sensing magnetic directions

in birds: radical pair processes involving cryptochrome', *Biosensors*, 4: 221-42.

46 Wiltschko, W., Traudt, J., Güntürkün, O., Prior, H., and Wiltschko, R. (2002), 'Lateralization of magnetic compass information in a migratory bird', *Nature*, 419: 467-70.

47 Stapput, K., Güntürkün, O., Hoffman, K.-P., Wiltschko, R., and Wiltschko, W. (2010), 'Magnetoreception of directional information in birds requires non-degraded vision', *Current Biology*, 20: 1259-62.

48 Falkenberg, G., Fleissner, G., Schuchardt, K., Kuenbacher, M., Thalau, P., Mourtisen, H., Heyers, D., Wallenreuther, G., and Fleissner, G. (2010), 'Avian magnetoreception: elaborate iron mineral containing dendrites in the upper beak seem to be a common feature of birds', *PlosOne*, 5: e9231.

49 Wiltschko, R., Gehring, D., Denzau, S., Güntüerkün, O., and Wiltschko, R. (2010), 'Interaction of magnetite receptors of the beak with the visual system underlying fixed direction responses in birds', *Frontiers in Zoology*, 7: 24.

50 Jenni-Eiermann, S., Jenni, L., Smith, S., and Costantini, D. (2014), 'Oxidative stress in endurance flight: an unconsidered factor in bird migration', *PlosOne*, 9: e97650.

51 http://blx.bto.org/birdfacts/results/bob10990.htm [accessed 25/8/15]

52 Snow, B., and Snow, D. (1988), *Birds and Berries: a study of an ecological interaction*.

53 Herrera, C. M. (1998), 'Long-term dynamics of Mediterranean Frugivorous birds and fleshy fruits: a 12-yr study', *Ecological Monographs*, 68: 511-38.

54 Herrera, C. M. (1977), 'Ecología alimentica del petirrojo (*Erithacus rubecula*) durante su invernada en encinares del Sur de España', *Doñana, Acta Vertebrata*, 4: 35-59.

55 Thomas, R. J., Székely, T., Cuthill, I. C., Harper, D. G. C., Newson, S. E., Frayling, T. D., and Wallis, P. D. (2002), 'Eye size in birds and the timing of song at dawn', *Proceedings of the Royal Society*, B 269: 831-7.

56 East, M. (1980), 'Sex differences and the effect of temperature on the foraging of Robins *Erithacus rubecula*', *Ibis*, 122: 517-20; East, M. (1982), 'Time-budgeting by European Robins *Erithacus rubecula*: inter and intrasexual comparisons during autumn, winter and early spring', *Ornis Scandinavica*, 13: 85-93; Harper, D. G. C. (1989), 'Individual territories in the European Robin', *Acta Ornithologica Internationalis* XIX, 2: 2355-633; Tobias, J. (1997), 'Food availability as a determinant of pairing behaviour in the European Robin', *Journal of Animal Ecology*, 66: 629-39.

57 Mead, C. J. (1982), 'Ringed birds killed by cats', *Mammal Review*, 12: 183-6.

58 Baker, P. J, Bentley, A. J., Ansell, R. J., and Harris, S. (2005), 'Impact of predation by domestic cats *Felis catus* in an urban area', *Mammal Review*, 35: 302-12.

59 Cuadrado, M. (1997), 'Why are migrant Robins (*Erithacus rubecula*) territorial in winter?: the importance of the anti-predatory behaviour', *Ethology, Ecology and Evolution*, 9: 77-88.

60 Johnstone, I. (1988), 'Territory structure of the Robin *Erithacus rubecula* outside breeding season', *Ibis*, 140: 244-51.

61 Examples: Hoelzel, A. R. (1989), 'Territorial behaviour of the Robin *Erithacus rubecula*: the importance of vegetation density', *Ibis*, 131: 432-6; Fennessy, G. J., and Kelly, T. C. (2006), 'Breeding densities of Robin *Erithacus rubecula* in different habitats: the importance of hedgerow structure', *Bird Study*, 53: 97-104.

62 Chantrey, D. F., and Workman, L. (1984), 'Song and plumage effects on aggressive display by the European Robin *Erithacus rubecula*', *Ibis*, 126: 366-71.

63 Mason, C. F. (1995), *The Blackcap*.

64 Grey, E. (1927), *The Charm of Birds*.

65 e.g.: Maynard Smith, J., and Harper, D. G. C. (2003), *Animal Signals*.

66 Fuller, R. J. (1995), *Bird Life of Woodland and Forest*.

67 Wesolowski, T., and Fuller, R. J. (2012), 'Spatial variation and temporal shifts in habitat use by birds on a European scale', in Fuller, R. J. (ed.), *Birds and Habitat Relationships in Changing Landscapes*, pp. 63-92.

68 Beven, G. (1976), 'Changes in breeding bird populations of an oakwood on Bookham Common, Surrey, over twenty-seven years', *London Naturalist*, 55: 23-42.

REFERENCES IN
THE LIFE OF 'THE LIFE OF THE ROBIN'

Burkitt, J. P. (1924-6), 'A study of the robin by means of marked birds',
    British Birds, 17: 294-303; 18: 97-103, 250-257; 19: 120-124; 20: 91-
    101.

East, M. (1981), 'Aspects of courtship and parental care of the
    European Robin *Erithacus rubecula*', *Ornis Scandinavica*, 12: 230-
    9.

Fox, A. D., and Beasley, P. D. L. (2010), 'David Lack and the birth of
    radar ornithology', *Archives of Natural History*, 37: 325-32.

Harper, D. G. C. (1985), 'Pairing strategies and mate choice in
    female robins *Erithacus rubecula*', *Animal Behaviour*, 33: 862-75.

Lack, A. (2007), *Redbreast: the Robin in Life and Literature*.

Lack, D. (1939), 'The behaviour of the robin. Part I: The life history, with
    special reference to aggressive behaviour, sexual
    behaviour, and territory. Part II: A partial analysis of aggressive
    and recognitional behaviour', *Proceedings of the Zoological Society
    of London*, 109: 169-219.

Lack, D. (1940a), 'The behaviour of the robin. Population changes over
    four years', *Ibis*, (14) 4: 299-324.

Lack, D. (1940b), 'Observations on captive robins', *British Birds*, 33: 262-
    70.

Lack, D. (1943), *The Life of the Robin*.

Lack, D. (1947), *Darwin's Finches*.

Lack, D. (1950), *Robin Redbreast*.

Lack, D. (1956), *Swifts in a Tower*.

Lack, D. (1973), 'My life as an amateur ornithologist', *Ibis*, 115: 421-31.

Lack, D. (1976), *Island Biology. Illustrated by the Land Birds of
    Jamaica*.

Lack, D., and Lack, H. L. (1933), 'Territory reviewed', *British Birds*, 27:
    179-99.

Lack, P. (2001), 'The legacy and significance of *The Life of the Robin* by
    David Lack', *British Wildlife*, 13: 95-100.

Nice, M. M. (1937), 'Studies in the life history of the song sparrow. I',
    Transactions of the Linnæan Society, New York, 4:
    1-247.

Snow, D. W. (1997), 'Comment', *Ibis*, 139: 572-5.

Tucker, B. W. (= B. W. T.) (1943), 'Review of *The Life of the Robin*',
    British Birds, 37: 120.

FURTHER READING AND RESOURCES

Clement, P., and Rose, C. (2015), *Robins and Chats.*

Collar, N. J. (2005), 'Family Turdidæ (Thrushes)', in del Hoyo, J., Elliott, A., and Christie, D. A., *Handbook of the Birds of the World. Vol. 10. Cuckooshrikes to Thrushes*, pp. 514-807.

Cramp, S. (1988), *Handbook of Birds of the Western Palearctic. Vol. 5. Tyrant Flycatchers to Thrushes*, pp. 596-616.

Golley, M., and Moss, S. (2011), *The Complete Garden Bird Book.*

Lack, A. (2007), *Redbreast: The Robin in life and literature.*

Mead, C. (1984), *Robin.*

Reid, M., King, M., and Allsop, J. (1995), *The Robin.*

Taylor, M. (2015), *Robin.*

BTO (British Trust for Ornithology), www.bto.org.

RSPB (Royal Society for the Protection of Birds), www.rspb.org.uk.

INDEX

This edition first published 2016 by
Pallas Athene (Publishers) Ltd,
Studio 11A, Archway Studios,
25-27 Bickerton Road,
London N19 5JT

Editor: Alexander Fyjis-Walker
Editorial assistance: Patrick Davies and Anaïs Métais
Special thanks to Peter Lack, Barbara Fyjis-Walker
    and Susan Lendrum

First edition published by H. F. Witherby, London 1943
Second edition H. F. Witherby, London 1946
Third edition Pelican Books, Harmondsworth 1953
Fourth edition H. F. & G. Witherby, London 1965,
    reprinted 1976
Reissued in Fontana New Naturalist series 1970, 1972
This edition reprinted 2016, 2017

www.pallasathene.co.uk

 @pallas_books   @PallasAtheneBooks

 @Pallas_books   @Pallasathene0

ISBN 978 1 84368 130 4